Engineering Chemistry

Engineering Chemistry

Hyon Carl

New York

Published by NY Research Press
118-35 Queens Blvd., Suite 400,
Forest Hills, NY 11375, USA
www.nyresearchpress.com

Engineering Chemistry
Hyon Carl

International Standard Book Number: 978-1-64725-428-5 (Hardback)

This book contains information obtained from authentic and highly regarded sources. All chapters are published with permission under the Creative Commons Attribution Share Alike License or equivalent. A wide variety of references are listed. Permissions and sources are indicated; for detailed attributions, please refer to the permissions page. Reasonable efforts have been made to publish reliable data and information, but the authors, editors and publisher cannot assume any responsibility for the validity of all materials or the consequences of their use.

Trademark Notice: Registered trademark of products or corporate names are used only for explanation and identification without intent to infringe.

Cataloging-in-Publication Data

Engineering chemistry / Hyon Carl.
    p. cm.
Includes bibliographical references and index.
ISBN 978-1-64725-428-5
1. Chemistry, Technical. 2. Chemical engineering. I. Carl, Hyon.
TP145 .E54 2023
660--dc23

Contents

Preface

Engineering chemistry is one of the branches of engineering involved in the integrated study of organic chemistry, analytical chemistry and electrochemistry. It deals with the operation and design of chemical plants and methods to improve production. This branch employs the principles of various fields such as chemistry, physics, mathematics, biology and economics, in order to efficiently use, produce, design, transport and transform energy and materials. The field of engineering chemistry focuses on developing economic ways of using materials and energy. It combines chemistry and engineering to convert raw materials into usable products, such as medicine and petrochemicals. Engineering chemistry encompasses several aspects of plant design and operation, including safety and hazard assessments, process design and analysis, modeling, control engineering, chemical reaction engineering, nuclear engineering, biological engineering, construction specification, and operating instructions. The objective of this book is to give a general view of the different areas of engineering chemistry as well as its applications. It aims to equip students and researchers with the advanced topics and upcoming concepts in this area.

The researches compiled throughout the book are authentic and of high quality, combining several disciplines and from very diverse regions from around the world. Drawing on the contributions of many researchers from diverse countries, the book's objective is to provide the readers with the latest achievements in the area of research. This book will surely be a source of knowledge to all interested and researching the field.

In the end, I would like to express my deep sense of gratitude to all the authors for meeting the set deadlines in completing and submitting their research chapters. I would also like to thank the publisher for the support offered to us throughout the course of the book. Finally, I extend my sincere thanks to my family for being a constant source of inspiration and encouragement.

Hyon Carl

High Polymers and Plastics

1.1 Polymerization: Mechanism, Methods and Properties

Polymers

The polymers are macro molecules which is formed by repeated linking of small molecules called monomers.

Eg: Nylon, PVC, Polyethylene.

Monomers

Monomers are micro molecule which combines with each other to form a polymer.

Eg: $CH_2 = CH_2$

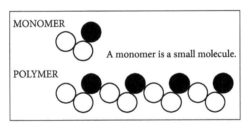

Structure of monomer and polymer.

Classification of Polymers

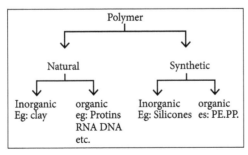

They are classified into two types:

- Natural polymers.

- Synthetic polymers.

Types

- Organic polymer.

- Element organic.

- Inorganic polymers.

Natural and Synthetic

Natural Polymers

The polymers obtained from nature (plants and animals are called natural polymers).

Example:

- Starch: It is a polymer of glucose. It is a chief food reserve of plants.

- Cellulose: It is a polymer of glucose. It is a chief structural material of the plants.

  Both Starch and cellulose are produced by plant during photosynthesis.

- National rubber: It is a polymer of unsaturated hydrocarbon, 2-methyl-1, 3 butadiene, called isoprene. It is obtained from latex of rubber trees.

$$nCH_2 = C\text{-}CH=CH_2 \xrightarrow{\text{polymerisation}}$$
$$\underset{CH_3}{|}$$
Isoprene

$$\underset{CH_3}{\overline{|}} +CH_2\text{-}C=CH\text{-}CH_2 \xrightarrow{}_n$$

Polyisoprene (NR)

Synthetic Polymers

The polymers which are prepared in the laboratories are called synthetic polymers. These are also called man-made polymers. Example: PVC, poly ethylene, Teflon, Bakelite etc.

Types

1. Organic polymer: These are polymers containing hydrogen, oxygen, nitrogen and

sulfur and halogen atoms apart from carbon atoms. Example: PVC (Polyvinyl chloride), Epoxy polymer, polyurethane.

2. Element organic: These are polymers composed of carbon atoms and heteroatoms (like N, S & O). The main chain consists of carbon atoms and whose side groups contain hetero atoms linked directly to the 'c' atoms in the chain. Example: Polysiloxane.

3. Inorganic polymers: These are polymer that contain no carbon atoms. The chains of these polymers are composed of different atoms joined by chemical bonds. Example: Polysilicon dioxide and polyphosphoric acid.

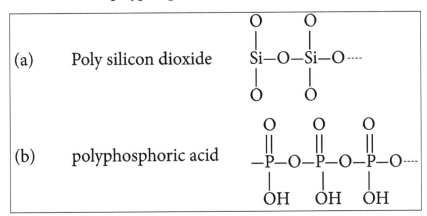

Polymerization - Degree of Polymerization

Polymerization is the process of converting small organic molecules into high molecular weight molecules either by addition or by condensation reaction. The small molecules are called monomers and the products are called polymers.

Degree of Polymerization

The number of repeating units present in the polymer chain is called the degree of polymerization. Degree of polymerization is given by,

$$\text{Degree of polymerization}(n) = \frac{\text{Molecular weight of the polymeric network}}{\text{Molecular weight of the repeating unit}}$$

Example:

$$5CH_2 = CH_2 \rightarrow -CH_2-CH_2-CH_2-CH_2-CH_2-CH_2-CH_2-CH_2-CH_2-CH_2$$

Here, five repeating units are present in the polymer chain.

So, the degree of polymerization = 5.

Types of Polymerization

There are two types of polymerization:

- Addition polymerization.

- Condensation polymerization.

Polymerization can proceed according to two different mechanisms referred to as chain-growth and step-growth polymerization.

Chain-growth polymerization is also called addition polymerization. This reaction occurs by successive addition of monomer molecules to the reactive end of a growing polymer chain.

In step-growth polymerization reaction between simple polar groups containing monomers with the formation of polymer and elimination of small molecules like H_2O, HCl, etc occurs.

Addition/chain Polymerization	Condensation / step polymerization
Longer reaction times give higher yield, but have a little effect on molecular weight.	Longer reaction times are essential to obtain high molecular weight.
Number of monomeric units decreases steadily through out the reaction.	Monomers disappear at the early stage of reaction.
High molecular weight polymer is formed at once.	Molecular weight of the polymer rise steadily throughout the reaction.
Molecular weight of the polymer is an integral multiple of molecular weight or monomer.	Molecular weight of the polymer need not be an integral multiple of monomer.
The monomer must have at least one multiple bond Eg: Ethylene $CH_2 = CH_2$.	The monomer must have at least two identical or different functional group Eg: Glycol $CH_2 - OH$ $\mid$ $CH_2 - OH$.
Monomers add on to give a polymer and no other by product is formed.	Monomers condense to give a polymer and by-products such as H_2O CH_3OH are formed.
Homo chain polymer is obtained.	Hetero-chain polymer is obtained.
Thermoplastics are produced. Eg: PVC etc.	Thermosetting plastic are produced. Eg: Urea - formaldehyde.

Additional Polymerization and the Mechanism of Free Radical Polymerization

It is a reaction that yields a polymer, which is an exact multiple of the original monomeric molecule usually contains one or more double bonds. In this addition polymerization there is no elimination of any molecule like HCL, H_2O etc.

Example:

- PVC is produced from Vinyl Chloride.

$$nCH_2= \underset{\underset{Cl}{|}}{CH} \xrightarrow{\text{Heat/Pressure}} n...+ (CH_2\text{-}\underset{\underset{Cl}{|}}{CH})_n$$

The mechanism of addition polymerization can be explained by one of the following types:

- Free radical mechanism.

- Ionic mechanism.

All the above mechanism occur in three major steps namely:

- Initiation.

- Propagation.

- Termination.

1. Initiation

It is considered to involve two reactions:

(a) First reaction involves production of free radicals by homolytic dissociation of an initiator (or catalyst) to yield a pair of free radicals (R*).

$$I \rightarrow 2R*$$

Where,

I = Initiator and R = Free Radicals

Commonly used thermal initiators:

Thermal initiator is a substance used to produce free radicals by homolytic dissociation at high temperature.

$$\underset{\text{Acetyl peroxide}}{CH_3COO\text{-}OOCCH_3} \xrightarrow{70\text{-}90°C} \underset{\text{Free radicals}}{2(CH_3COO)(or)2R°}$$

(b) Second reaction involves addition of this free radical to the first monomers to produce chain initiating species.

$$R + CH_2 = \overset{\overset{\displaystyle H}{|}}{\underset{\underset{\displaystyle Y}{|}}{C}} \longrightarrow R - CH_2 = \overset{\overset{\displaystyle H}{|}}{\underset{\underset{\displaystyle Y}{|}}{C}}$$

(Free radical)   First   Chain initating
                  monomer   species

2. Propagation

It involves the growth of chain initiation species by successive addition of large number of monomers.

$$R-CH_2 = \overset{\overset{\displaystyle H}{|}}{\underset{\underset{\displaystyle Y}{|}}{\overset{\bullet}{C}}} + nCH_2 = \overset{\overset{\displaystyle H}{|}}{\underset{\underset{\displaystyle Y}{|}}{C}} \longrightarrow R + (CH_2 - \overset{\overset{\displaystyle H}{|}}{\underset{\underset{\displaystyle Y}{|}}{C}} - CH_2)_n \overset{\overset{\displaystyle H}{|}}{\underset{\underset{\displaystyle Y}{|}}{\overset{\bullet}{C}}}$$

Growing chain (living polymer

The growing chain of the polymer is known as living polymer.

3. Termination

Termination of the growing chain of polymer may occur either by coupling reaction (or) disproportionation.

(a) Coupling or Combination

It involves coupling of free radical of one chain end to another free radical forming macromolecule so called as dead polymer.

$$R-CH_2 - \overset{\overset{\displaystyle H}{|}}{\underset{\underset{\displaystyle Y}{|}}{\overset{\bullet}{C}}} + \overset{\overset{\displaystyle H}{|}}{\underset{\underset{\displaystyle Y}{|}}{\overset{\bullet}{C}}} - CH_2 - R \longrightarrow R - CH_2 - \overset{\overset{\displaystyle H}{|}}{\underset{\underset{\displaystyle Y}{|}}{C}} - \overset{\overset{\displaystyle H}{|}}{\underset{\underset{\displaystyle Y}{|}}{C}} - CH_2 - R$$

Macro molecule
(Dead polymer)

(b) Disproportionation

It involves transfer of a hydrogen atom of one radical center to another radical center forming two macro molecules one saturated and another unsaturated.

$$R - \overset{\overset{\displaystyle H}{|}}{\underset{\underset{\displaystyle H}{|}}{C}} - \overset{\overset{\displaystyle H}{|}}{\underset{\underset{\displaystyle Y}{|}}{\overset{\bullet}{C}}} + \overset{\overset{\displaystyle H}{|}}{\underset{\underset{\displaystyle Y}{|}}{\overset{\bullet}{C}}} - \overset{\overset{\displaystyle H}{|}}{\underset{\underset{\displaystyle Y}{|}}{C}} - R \longrightarrow R - \overset{\overset{\displaystyle H}{|}}{\underset{\underset{\displaystyle }{}}{C}} - \overset{\overset{\displaystyle H}{|}}{\underset{\underset{\displaystyle Y}{|}}{C}} + \; + H - \overset{\overset{\displaystyle H}{|}}{\underset{\underset{\displaystyle Y}{|}}{C}} - \overset{\overset{\displaystyle H}{|}}{\underset{\underset{\displaystyle H}{|}}{C}} - R$$

Unsaturated   Saturated

Anionic Polymerization

Addition polymers can also be made by chain reactions that proceed through intermediates that carry either a negative or positive charge.

When the chain reaction is initiated and carried by negatively charged intermediates, the reaction is known as anionic polymerization. Like free-radical polymerizations these chain reactions take place via chain-initiation, chain-propagation and chain-termination steps.

The reaction is initiated by a Grignard reagent or alkyl lithium reagent, which can be thought of a source of a negatively charged CH_3- or CH_3CH_2- ion.

$$H-\underset{\underset{H}{|}}{\overset{\overset{H}{|}}{C}}-\underset{\underset{H}{|}}{\overset{\overset{H}{|}}{C}}-Li \longleftrightarrow H-\underset{\underset{H}{|}}{\overset{\overset{H}{|}}{C}}-\underset{\underset{H}{|}}{\overset{\overset{H}{|}}{C}}:- \quad Li^+$$

The CH_3- or CH_3CH_2- ion from one of these metal alkyls can attack an alkene to form a carbon-carbon bond.

$$H-\overset{\overset{H}{|}}{\underset{\underset{H}{|}}{C}}-\overset{\overset{H}{|}}{\underset{\underset{H}{|}}{C}}:- \quad + \quad \overset{H}{\underset{H}{}}C=C\overset{H}{\underset{X}{}} \longrightarrow H-\overset{\overset{H}{|}}{\underset{\underset{H}{|}}{C}}-\overset{\overset{H}{|}}{\underset{\underset{H}{|}}{C}}-\overset{\overset{H}{|}}{\underset{\underset{H}{|}}{C}}-\overset{\overset{H}{|}}{\underset{\underset{X}{|}}{C}}:-$$

The product of this chain-initiation reaction is a new carbanion that can attack another alkene in a chain-propagation step.

$$H-\overset{\overset{H}{|}}{\underset{\underset{H}{|}}{C}}-\overset{\overset{H}{|}}{\underset{\underset{H}{|}}{C}}-\overset{\overset{H}{|}}{\underset{\underset{H}{|}}{C}}-\overset{\overset{H}{|}}{\underset{\underset{X}{|}}{C}}:- \quad + \quad \overset{H}{\underset{H}{}}C=C\overset{H}{\underset{X}{}} \longrightarrow H-\overset{\overset{H}{|}}{\underset{\underset{H}{|}}{C}}-\overset{\overset{H}{|}}{\underset{\underset{H}{|}}{C}}-\overset{\overset{H}{|}}{\underset{\underset{H}{|}}{C}}-\overset{\overset{H}{|}}{\underset{\underset{X}{|}}{C}}-\overset{\overset{H}{|}}{\underset{\underset{H}{|}}{C}}-\overset{\overset{H}{|}}{\underset{\underset{X}{|}}{C}}:-$$

The chain reaction is terminated when the carbanion reacts with traces of water in the solvent in which the reaction is run.

$$CH_3CH_2(CH_2\underset{\underset{X}{|}}{CH})_nCH_2\underset{\underset{X}{|}}{CH}:- \quad + \quad H_2O \longrightarrow CH_3CH_2(CH_2\underset{\underset{X}{|}}{CH})_nCH_2\underset{\underset{X}{|}}{CH_2} \quad + \quad OH^-$$

Cationic Polymerization

The intermediate that carries the chain reaction during polymerization can also be a positive ion or cation. In this case, the cationic polymerization reaction is initiated by adding a strong acid to an alkene to form a carbonation.

$$CH_2 = \underset{\underset{X}{|}}{CH} + H^+ \longrightarrow CH_3\underset{\underset{X}{|}}{CH^+}$$

The ion produced in this reaction adds monomers to produce a growing polymer chain.

$$CH_3CH^+ + CH_2 = CH \longrightarrow CH_3CHCH_2CH^+$$
$$\qquad | \qquad\qquad | \qquad\qquad\qquad | \qquad |$$
$$\qquad X \qquad\qquad X \qquad\qquad\qquad X \qquad X$$

The chain reaction is terminated when the carbonium ion reacts with water that contaminates the solvent in which the polymerization is run.

$$CH_3CH(CH_2CH)_nCH_2CH^+ + H_2O \longrightarrow CH_3CH(CH_2CH)_nCHOH + H^+$$
$$\qquad | \qquad\quad | \qquad\quad | \qquad\qquad\qquad | \qquad\quad | \qquad\quad |$$
$$\qquad X \qquad\quad X \qquad\quad X \qquad\qquad\qquad X \qquad\quad X \qquad\quad X$$

Stereoregular Polymer

The IUPAC commission on macromolecular nomenclature defines a stereoregular polymer as "a regular polymer whose molecules can be described by only one species of stereo repeating unit in a single sequential arrangement". A stereo repeating unit is further defined as "a configurational repeating unit having defined configuration at all sites of stereoisomerism in the main chain of a polymer molecule". Tactic polymers are considered to be those which exhibit a regular pattern of configurations for at least one type of stereoisomerism sites in the main chain.

According to these nomenclature proposals, one can have tactic stereoregular polymers and tactic non-stereoregular polymers, according to whether every type or only one type of stereoisomerism site in the main chain is in a configurationally ordered sequence. Cases in which steric order exists only for stereoisomerism sites in the side chains, the main chain being non-tactic (not containing stereoisomerism sites) or atactic are not considered.

Methods of Polymerization

The following methods are generally used for the polymerization reaction:

- Bulk polymerization.

- Solution polymerization.

- Suspension polymerization.

- Emulsion polymerization.

1. Bulk Polymerization

Bulk polymerization is the simplest method of polymerization. The monomer is taken in a flask as a liquid form and the initiator chain transfer agent are dissolved in it. The flask is placed in thermostat under constant agitation and heated.

$$\text{Monomer} + \text{initiator} + \text{chain transfer agent polymer}$$

The reaction is slow but becomes fast as the temperature rises. After a known period of time, the whole content is poured into a methanol (non solvent) and the polymer is precipitated out.

Polystyrene, PVC, PMA are prepared by this method.

Advantages

- It has high purity in the polymer reaction.

Disadvantages

- Polymerization is highly exothermic.

- During the polymerization viscosity of the medium increases mixing and control of heat is difficult.

Applications

- Low molecular weight polymers, can be used as adhesives, plasticizers and lubricant additives.

- Polymer obtained by this method are used in casting formulations.

2. Emulsion Polymerization

Emulsion Polymerization is used for water insoluble monomers and water soluble initiator like potassium persulphate.

$$\text{Monomer} + \text{Initiator} + \text{Surfactant} \rightarrow \text{Polymer}$$

Advantages

- The rate of polymerization is high.

- High molecular weight polymer can be obtained.

Disadvantages

- Polymer needs purification.

- Requires rapid agitation.

Applications

- Emulsion polymerization is used in large scale production.

- Manufacturing tacky polymers like butadiene and chloroprene.

3. Solution Polymerization

In solution polymerization the monomers initiator and the chain transfer agent are taken in a flask and dissolved in an inert solvent. The whole mixture is kept under constant agitation. After required time, the polymer produced is precipitated by pouring it in a suitable non-solvent.

$$\text{Monomer + Initiator + chain transfer agent} \rightarrow \text{Polymer}$$

The solvent helps to control heat and reduce viscosity built up (e.g.: Polacrylic acid, polyisobutylence prepared in this method).

Advantages

- Viscosity built up is negligible.

- The mixture can be agitated easily.

Disadvantages

- The removal of last traces of solvent is difficult.

- This polymerization requires solvent recovery and recycling.

- It is difficult to get very high molecular weight polymer.

Application

- Polymer is in solution form it can be directly used as adhesives and coatings.

4. Suspension Polymerization

Suspension Polymerization is used only for water insoluble monomers. This polymerization reaction is carried out in heterogeneous system. At the end of the Polymerization, polymer is separated out as spherical beads or pearls. This method is also called pearl Polymerization.

$$\text{Monomer + initiator + suspending agent} \rightarrow \text{Polymer}$$

Examples: polystyrene

Advantage

- Since water is used as a solvent, this method is more economical.

- Products obtained is highly pure.

- Isolation of product is very easy.

- Efficient heat control.

Disadvantages

Control of particle size is difficult.

Application

Polystyrene beads are used as ion exchangers.

Properties of Polymers

The following are the important properties considered for the polymers:

- Glass transition temperature-$\left(T_g\right)$.

- Weight - average molecular mass $\left(\overline{Mw}\right)$.

- Tacticity.

- Number average.

- Polydispersity index.

Physical Properties of Polymers

- Strength: It depends upon the magnitude of force of attraction between polymeric chains. Two types (a) Primary or chemical bond and (b) Secondary or intermolecular forces (van-der Wall force or hydrogen bonding).

- Deformation: Deformation is the slipping of one chain over the other (on the application of heat or pressure or both) or stretching and recoverance of original shape of the polymeric chains (after the removal of stress).

Types

- Plastic Deformation (Plasticity).

- Elastic Deformation (Elasticity).

3. Chemical Resistance and Solubility

Chemical attack is internal, causing softening, swelling and loss of strength of polymer. The chemical nature of monomeric units and their molecular arrangement determines the chemical resistance of the polymer.

Polymers having polar groups (-OH, −COOH or Cl) swollen or even dissolved in polar solvents whereas polymers having nonpolar groups $\left(-C_6H_5 \text{ or } -CH_3\right)$ swollen or even dissolved in non-polar solvents.

Polymers of more aliphatic characters are more soluble in aliphatic solvents whereas polymers of more aromatic character are more soluble in aromatic solvents.

The tendency to solubility or swell of polymers decreases with the increase in molecular weight or chain length of polymer. High molecular weight polymers on dissolving yield solutions of high viscosities.

4. Physical state of Polymers

An amorphous state is characterized by a completely random, irregular and dissymmetrical arrangement of polymeric chains. e.g., thermosetting, rubber.

A crystalline state is characterized by a completely regular, symmetrical and ordered arrangement of polymeric chains with uniaxial orientation. e.g., fibers.

A semi-crystalline state consists of crystalline region called as crystallites (ordered arrangement of polymeric chains) embedded in an amorphous matrix. e.g., thermoplastic.

5. Effect of Heat (Glass Transition Temperature T_g)

Mechanical Properties of Polymers

1. Hardness: The ability of a polymer to resist scratching, cutting or penetration, abrasion. It is measured by its ability to absorb energy under impact loads. Hardness is associated with strength. It is closely associated with material structure, composition and other mechanical properties.

2. Toughness: It is the amount of energy a polymer can absorb before actual fracture or failure takes place. The ability of a polymer to withstand shock and vibrations. It is related to impact strength which is the resistance to breakage under high velocity impact conditions, i.e., resistance to shock loading. It is the ability of a polymer to withstand both plastic and elastic deformation.

3. Stiffness: The resistance of a polymer to elastic deformation, i.e., a polymer which suffers slight deformation under load has a high degree of stiffness. Flexibility is the opposite of stiffness.

4. Density: Relative Density is the mass of the polymer with the mass of equal volume of a specific (reference) substance (water). Density is frequently measured as a quality control parameter.

A specimen with smooth surfaces from crevices and dust is weighed in air (W_1) and then in freshly boiled water (W_2), then,

$$\rho_{polymer} = \frac{W_1}{W_1 - W_2} \rho_{water}$$

5. Tensile Strength: The strength of a polymer is its capacity to withstand destruction under the action of loads. It determines the ability of a polymer to withstand stress without failure. The Tensile strength or ultimate strength is the stress corresponding to the maximum load reached before rupturing the polymer.

$$\text{Tensile strength or Stress} = \frac{\text{Force or Maximum Load}}{\text{Area of Cross Section}}$$

6. Abrasion Resistance: It is defined as the ability of a polymer to withstand mechanical action (such as scrapping, rubbing, erosion) that tends progressively to remove material from its surface.

Abrasion is closely related to frictional force, the load and true area of contact, an increase in any one of the three results in greater abrasion or wear. Abrasion process also creates oxidation on the surface from the buildup of localized high temperatures.

7. Resilience: It is the capacity of a polymer to absorb energy elastically. Resilience gives capacity to the polymer to bear shocks and vibrations. When a body is loaded, it changes its dimension and on the removal of the load it regains its original dimensions.

In fact, the polymer behaves perfectly like a spring, so long as it remains loaded, it has stored energy in itself, on removal of the load, the energy stored is given off exactly as in a spring when the load is removed.

8. Wear and Tear: It occurs when there is a steady rate of increase in the use of polymers in bearing applications and in situations where there is sliding contact e.g. gears, piston rings, seals, cams, etc. Wear and tear is characterized by the fine particles of polymer being removed from the surface or the polymer becomes overheated to the extent where large troughs of melted polymer are removed.

The wear and tear of polymers is extremely complex subjects which depends on the nature of the application and the properties of the material. It is characterized by adhesion and deformation which results in frictional forces that are not proportional to load but rather to speed. The mechanism of wear and tear is complex, the relative rates may change depending on specific circumstance.

1.2 Plastics

Plastics are products of polymers. Polymers are resins which could be molded into different shapes by using heat and pressure.

Plastics as Engineering Materials

Engineering plastics are those which possess physical properties enabling them to perform for prolonged use in structural applications, over a wide temperature range, under mechanical stress and in difficult chemical and physical environments.

Engineering plastics are most frequently thought of as the acetals, nylons, fluorocarbons, phenolics, polycarbonate and polyphenylene oxide. These are indeed engineering materials and for such applications are usually used in relatively small amounts, in comparison with the nonengineering plastics which are used in commodity quantities.

Advantages

For most engineering applications, plastics are considered to be competitive primarily with metals, although there is considerable competition among the plastics themselves. In comparison with metals, plastics have certain properties which are generally considered to be advantageous for engineering applications. For the most part, plastics have better chemical and moisture resistance and are more resistant to shock and vibration than metals. They are lighter in weight and usually either transparent or at least translucent in thin sections.

They have the advantage of absorbing sound and vibration and some possess greater wear and abrasion resistance than metals. Some of them, as in the case of nylons, for example, are self-lubricating. Significantly, one of the most important characteristics of plastics is that they usually are easier to fabricate than metals.

Some plastics can be plated but perhaps an even more important property is that plastics can be pigmented in a wide variety of colors. Finally, because of their lighter weight, giving many of them an advantage in cost per cubic inch with respect to metals and because they are usually easier to fabricate, finished parts made of plastics are frequently less costly than those made of metal.

Disadvantages

Plastics are not as strong as metals. Generally, they possess lower heat resistance and most of them are flammable. They are characterized by a much larger thermo expansion, are often less ductile than metals and most are much more subject to embrittlement at low working temperatures and unfortunately, for many applications for which plastics are used or desired to be used, they are more susceptible to creep.

Plastics are softer than metals and some of them, when they absorb water or solvents, will change in dimension, quite a negative feature when these materials are used for gears and other close tolerance parts. In addition, most plastics are subject to degradation by ultraviolet light. Finally, most plastics cost more than metals on a per-pound basis and some of them cost more than some metals on a per cubic inch basis.

Unfilled plastics are thermal and electrical non-conductors and they are subject to deformation by heat and/or pressure. Fortunately, however, both for the engineer and the plastics manufacturer, plastics can be modified by the addition of other materials. In addition, fire retardants, ultraviolet absorbers and other additives can be incorporated into plastics to improve their properties and fillers and reinforcing materials can be incorporated into the plastics to impart many advantages.

Thermoplastics or Thermosoftening Plastics

A thermoplastic is a material which becomes soft when heated and hard when cooled. Thermoplastic materials can be cooled and heated several times.

These show reversible change on heating i.e. they soften on heating but regain their original properties on cooling.

They gain or lose hardness with rise or fall in temperature. Their chemical nature does not get affected even on repeated heating and cooling, i.e. the changes are more of physical nature.

If these resins are softened, they retain their softness at that temperature. These resins can be reclaimed from waste and they are soft, weak and less brittle. The method usually used to manufacture polymers is addition polymerization. They are generally long chain linear polymers with occasional or no cross linking. They are soluble in specific organic solvents.

Examples of this class of resins are cellulose nitrate, polyacrylates, ethyl cellulose, polyvinyl resins, styrene or polystyrene resins, polyamides (nylons) polyethers, polypropylene, polyethylene etc.

Thermosetting Plastics

Thermosetting Plastics cannot be remolded because all the polymer chains are linked by strong covalent bond.

These are the polymers, which on heating change irreversibly into hard and rigid materials. The melt of this polymer when set into a mould to form an article, is almost a permanent set. On reheating the article does not soften again, thus inhibiting irreversibility.

They are hence known as thermo hardening plastics or permanent setting resins and during moulding acquire three dimensional cross linked structure with strong covalent bonds.

On reheating, these bonds retain their strength and hence such a plastic does not soften on reheating.

Thermo setting plastics cannot be reclaimed from waste due to their irreversibility. They are hard, strong and brittle than thermo plastics. The method by which these are formed is called as condensation polymerisation.

They are insoluble in almost all organic solvents, due to their cross linked three dimensional structure. Phenol formaldehyde/bakelite, amino plastics, alkyl plastics, epoxy plastics, silicon plastics etc. are the best known examples of thermosets.

Compounding and Fabrication (4/5 Techniques)

Fabrication of Plastics

Plastic fabrication is the design, manufacture or assembly of the plastic products through any one out of the various methods that are present. Some manufacturers prefer fabrication of plastic over working with other materials due to the process's advantages in certain applications. Plastic's malleability and cost effectiveness could make it a versatile and durable material for a range of different products.

The primary processing of plastics involves first heating the material until it is liquid or semi-solid, forming it into the required shape and then solidifying it:

- By chemical curing in the case of thermosets.

- By simple cooling in the case of thermoplastics.

The fabrication and processing of plastics is highly complex involving many factors, viz- chemical, mechanical and physical playing an important role.

Coal, petroleum, cotton, wood, gas, air, salt and water are the basic raw materials, binders, fillers, plasticisers, colors, catalysts and lubricants are also added to plastics to provide them with different properties and to enhance their practical applicability.

Compounding or Blending

Compounding is a type of fabrication which can combine two or more plastics into an amalgam, before forming them into a single part. It involves mixing molten plastics to exact specifications and then forming them with a mold, die or other shaping tool.

Compounding is often used to improve the ease of processing a given material or to enhance the product performance. By combining the advantages and disadvantages of several types of plastic, the process can result in a unique material complementary to a specific application.

Some common types of plastic compounds include:

- Polymer fillers.
- Base resins.
- Pigment master batches.
- Blowing agents.
- Flame-retardants.
- Purge compounds.

1.3 Polyethene, PVC, Bakelite, Teflon and Polycarbonates

Polyethene

The liquid gases under high pressure is pumped into the heated pressure vessel maintained 150 to 25°C. By the catalytic effect of traces of the oxygen present ethylene is polymerized in to poly ethylene.

Ethylene-Ethylene bond yields polyethylene $(CH_2CH_2)_n$.

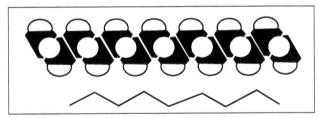

Structure of polyethylene.

Preparation

Polyethene is made by several methods by addition polymerization of ethene, which is principally produced by cracking of ethane and propane, naphtha and gas oil. A new plant is being constructed in Brazil for production of poly(ethene), from ethene, that is made from sugarcane via bio ethanol.

Low Density Poly(Ethene)

The process is operated under very high pressure (1000-3000 atm) at moderate temperatures (420-570 K) as may be predicted from reaction equation,

$$nC_2H_4 \longrightarrow \left[CH_2 - CH_2 \right]_n$$

$$\Delta H^{\ominus} = -92\,KJmol^{-1}$$

This is a radical polymerization process and an initiator, such as a small amount of oxygen and/or an organic peroxide is used.

Ethene (purity in excess of 99.9%) is compressed and passed into the reactor together with the initiator. The molten polyethene is removed, extruded and cut into granules. The Un-reacted ethene is recycled. The average polymer molecule contains 4000-40000 carbon atoms, with many short branches.

High Density Poly(Ethene)

Two types of catalyst are used principally in the manufacture of HDPE:

A Ziegler-Natta organometallic catalyst (titanium compounds with an aluminium alkyl).

- An inorganic compound, known as a Phillips-type catalyst. A well-known example is chromium(VI) oxide on silica, which is prepared by roasting a chromium(III) compound at ca 1000 K in oxygen and then storing prior to use, under nitrogen.

- HDPE is produced by three types of process. All operate at relatively low pressures (10-80 atm) in the presence of a Ziegler-Natta or inorganic catalyst. Typical temperatures range between 350-420 K. In all three processes hydrogen is mixed with the ethene to control the chain length of polymer.

1. Slurry Process (using Either CSTR (Continuous Stirred Tank Reactor i.e. CSTR or a Loop)

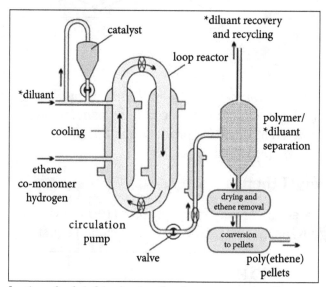

Production of poly(ethene) using the slurry process in a loop reactor.

The Ziegler-Natta catalyst, as granules, is mixed with a liquid hydrocarbon (for example, 2-methylpropane (isobutane) or hexane), which simply acts as a diluent. A mixture of hydrogen and ethene is passed under pressure into slurry and ethene is polymerized to HDPE. The reaction takes place in a large loop reactor with the mixture constantly stirred. On opening a valve, the product is released and the solvent is evaporated to leave the polymer, still containing the catalyst. The water vapour, on flowing with nitrogen through the polymer, reacts with the catalytic sites, destroying their activity. The residue of catalyst, titanium(IV) and aluminium oxides, remains mixed, in minute amounts, in the polymer.

2. Solution Process

The second method involves passing the ethene and hydrogen under pressure into a solution of the Ziegler-Natta catalyst in a hydrocarbon (a C_{10} or C_{12} alkane). The polymer is obtained in a similar way to the slurry method.

3. Gas Phase Process

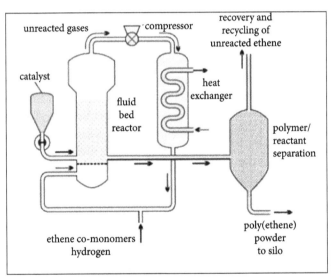

Low pressure gas-phase process.

A mixture of ethene and hydrogen is passed over a Phillips catalyst in a fixed bed reactor. The ethene polymerizes to form grains of the HDPE, suspended in the flowing gas, which pass out of the reactor when the valve is released.

The modern plants sometimes use two or more of the individual reactors in series (for example two or more slurry reactors or two gas phase reactors) each of which are under slightly different conditions, so that the properties of different products from the reactors are present in the resulting polymer mixture, leading to a broad or bimodal molecular mass distribution. This provides improved mechanical properties such as stiffness and toughness.

Linear Low Density Polyethene

Low density polyethene has many uses but the high pressure method of manufacture by which it is produced has high capital costs. However, an elegant technique has been developed, based on both Ziegler-Natta and inorganic catalysts to produce linear low density polyethene LLDPE, which has even improved properties over LDPE.

Any of the three processes, slurry, solution and gas phase, can be used when a Ziegler-Natta catalyst is chosen. The gas phase process is used when the inorganic catalyst is employed.

Small amounts of a co-monomer such as but-1-ene or hex-1-ene are added to the feedstock. The monomers are randomly polymerized and there are small branches made up of a few carbon atoms along the linear chains.

Metallocene Linear Low Density Polyethene

This polyethene, known as mLLDPE, is produced by the new family of catalysts, the metallocenes. Another name for this family is single site catalyst. The benefit is that mLLDPE is much more homogeneous in terms of molecular structure than classical LLDPE produced by the Ziegler-Natta catalysts.

Each catalyst is a single site catalyst which produces the same PE chain. Chemists have compared the structure of metallocenes to that of a sandwich. There is a transition metal (often zirconium or titanium) 'filling' a hole between layers of organic compounds.

The catalysts are even more specific than the original Ziegler-Natta and it is possible to control the polymer's molecular mass as well as its configuration. Either the slurry or solution processes are commonly used.

Poly(ethene) produced using a metallocene can be used as very thin film which has excellent optical properties and sealing performance, thus making them very effective for wrapping foods. The real advantage for metallocene catalysts are the enhanced mechanical properties of the films made out of mLLDPE.

Properties

- A rigid waxy solid white, translucent non-polar material.

- Chemically resistant to strong acids, alkalies and salt solutions.

- Good insulator of electricity.

- Swollen and permeable to most oils and organic solvents particularly to kerosene.

- Due to its high symmetrical structure the polyethylene crystallizes very easily.

- Polyethylene produced by high pressure process has a branched structure and therefore flexible and tough.

- Low pressure process results in a completely linear PE having high density and better chemical resistance.

Commercial PE is divided in to 3 types:

- Type I or low density PE (0.91-0.925g/cm_3).

- Type II or medium density PE(0.925 -0.940 g/cm_3).

- Type III or high density PE (0.941- 0.965 g /cm_3).

Applications

- For making high frequency insulator parts.

- Bottle caps.

- Flexible bottles.

- Kitchen and domestic appliances.

- Toys.

- Sheets for packing materials.

- Tubes pipes.

- Coated wires and cables.

PVC (Polyvinyl Chloride)

Polyvinyl chloride abbreviated as PVC is a thermoplastic resin. PVC has a unique amorphous structure with polar chlorine atoms in the molecular structure. Having chlorine atoms and the amorphous molecular structure are inseparably related.

Although plastics seem very similar in the daily use context, PVC has completely different features in terms of performance and functions as compared with olefin plastics which have only carbon and hydrogen atoms in their molecular structures.

Preparation

Polyvinyl chloride is prepared by polymerisation of monomer vinyl chloride

- About 80% of polymerisation includes suspension polymerisation.

- 12% emulsion polymerisation and 8% bulk polymerisation.

Process

The VCM and water are introduced into the reactor and a polymerization initiator, along with other additives. Reaction vessel is pressure tight to contain the VCM. The contents of the reaction vessel are continually mixed to maintain the suspension and ensure a uniform particle size of the PVC resin.

The reaction is exothermic and thus requires cooling. As the volume is reduced during the reaction (PVC is denser than VCM), water is continuously added to the mixture to maintain suspension. The polymerization of VCM is started by compounds called initiators that are mixed into droplets. These compounds break down to start the radical chain reaction.

Typical initiators include dioctanoyl peroxide and diacetyl peroxydicarbonate, both of which have fragile O-O bonds. Some initiators start the reaction rapidly but decay quickly and other initiators have the opposite effect. A combination of two different initiators is often used to give a uniform rate of the polymerization. After the polymer has grown by about 10x, the short polymer precipitates inside the droplet of the VCM and polymerization continues with the precipitated, solvent-swollen particles.

Once the reaction has run its course, the resulting PVC slurry is degassed and stripped to remove excess VCM, which is recycled. The polymer is then passed through a centrifuge to remove water. The slurry is further dried in a hot air bed and the resulting powder is sieved before storage or pelletization. Generally, the resulting PVC has a VCM content of less than 1 part per million.

Properties of PVC

- Weathering Stability: PVC is resistant to the aggressive environmental factors therefore it is a material of choice for roofing.

- Versatility: PVC can either be flexible or rigid.

- Fire Protection: PVC is a material that is resistant to ignition due to its chlorine content.

- Longevity: PVC products can last for more than 100 years.

- Hygiene: PVC is one of the material of choice for medical applications, particularly blood and plasma storage containers.

- Energy Recovery: PVC has high thermal power, when utilized in incinerators PVC provides power and heat for homes and industries and all that without having any impact on the environment.

- Barrier Properties: The PVC can be made impervious to the liquids, vapour and gases.

- Eco-efficiency: Only 43% of the PVC's content comes from oil (57% comes from salt), it therefore contributes to the preservation of that highly valuable natural resource.

- Recyclability: PVC is very recyclable, more so than many other plastics.

- Public Safety: PVC has often fallen under unfounded attempts so that today it is one of the best explored materials in the world due to serious scientific researches carried in order to disprove accusations.

- Economical Efficiency: The PVC is the cheapest of large-tonnage polymers providing many products with the best quality-price ratio.

Applications

- PVC is mainly used as an insulating material.

- It is used for making table clothes, rain coats, toys, tool handles, radio components, etc.

- It is used for making pipes, hoses, etc.

- It is used for making helmets, refrigerator components, etc.

- It is used in making cycle and automobile parts.

- It is used for sewerage pipes and other pipe applications where cost or vulnerability to corrosion limit the use of metal.

- With the addition of impact modifiers and stabilizers, it has become a popular material for window and door frames.

- By adding plasticizers, it can become flexible enough to be used in the cabling applications as a wire insulator.

Bakelite (Phenol-formaldehyde)

Principle

Phenol formaldehyde resins (PFs) are condensation polymers and are obtained by condensing phenol with the formaldehyde in the presence of an acidic or alkaline catalyst. They were first prepared by Backeland, an American Chemist who gave them the name as Bakelite. These are thermosetting polymers.

Preparation

Bakelite is the commercial name for polymer obtained by the polymerization of phenol and formaldehyde. This is one of the oldest polymers that was synthesised by man.

Phenol is made to react with formaldehyde. The condensation reaction of two reactants in a controlled acidic or basic medium results in the formation of ortho and para hydroxymethylphenols and their derivatives.

The reaction that takes place is as shown below,

Cross linked polymer-bakelite

When the phenol is used in excess and the reaction medium is made acidic, the product of the condensation reaction obtained is novolac. Whereas, when the quantity of formaldehyde taken exceeds the quantity of phenol in the reacting mixture and the reaction occurs in a basic medium, the condensation product obtained is known as resol.

Bakelite has a low electrical conductivity and high resistance to heat and chemicals. It is a thermosetting polymer and has high strength and retains its shape after moulding.

Properties

Phenol-formaldehyde resins having low degree of polymerization are soft. They possess excellent adhesive properties and are generally used as bonding glue for laminated wooden planks and in varnishes and lacquers.

Phenol-formaldehyde resins having high degree of polymerization are hard, rigid, scratch resistant and infusible. They are resistant to non-oxidising acids, salts and many organic solvents. They can withstand very high temperatures.

They act as excellent electrical insulators also:

- It does not conduct electricity.
- It is resistant to heat and is nonflammable.
- It is also resistant to chemical action.
- The dielectric constant of Bakelite ranges from 4.4 to 5.4.

Applications

- It is used for making TV cabinets, housing laminates, telephone components, decorative articles, bearings, electrical goods, etc.

- Bakelite is used as a substitute of porcelain and other opaque ceramic materials.

- It is used in the area of board and table top games.

- It is used in mounting of metal samples.

- Bakelite is very suitable for emerging electrical and automobile industries because of its extra ordinary resistance but also due to its thermal resistance.

- It is also used as an excellent adhesive.

Teflon

Teflon is a plastic like substance which is produced by polymerizing tetrafluoroethylene $(CF_2 = CF_2)$.

Poly Tetra Fluoro Ethylene(PTFE)

Polytetrafluroethylene is also known as teflon. It is a synthetic fluoro polymer of tetrafluoroethylene. It is a polymer which have numerous applications.

The IUPAC (International Union of Pure and Applied Chemistry) name of teflon is, Poly(1, 1, 2, 2-tetrafluoroethylene).

Formation of Polytetrafluoroethylene or Teflon or (PTFE)

It is formed when chloroform is treated with the hydrofluoric acid and antimony trifluoride.

$$\underset{\text{tetrafluoro ethylene}}{aCF_2 = CF_2} \rightarrow \underset{\text{teflon}}{\left(CF_2 - CF_2\right)_n}$$

Properties of Teflon or (PTFE)

- Teflon is one of the chemically inert substance and it is not affected by strong acids which are chemically harmful and even after by boiling aqua-regia.

- It has the property to be stable at high temperatures.

- It is a thermoplastic polymer that appears as a white solid at room temperature having a density of about 2200 kg/m³. It has a melting point of 600 K.(327 °C, 620 °F).

- It bears mechanical properties such that it can degrade gradually at temperatures above 194 K ($-79°C$, $-110°F$).

- PTFE mainly consist of carbon-fluorine bonds and it also gains the properties from the bonds created. Alkali metals and most highly reactive fluorinating agents are the only chemicals that can affect its property.

- It has a co-efficient of friction that is 0.05 to 0.10 which is the third-lowest of any known solid material. It has one of the proficient dielectric properties.

Applications of (PTFE)

- PTFE is usually used to coat in the non-stick frying pans as it has the ability to resist high temperatures.

- It is mostly used as a film interface patch for sports and medical applications, having a pressure-sensitive adhesive backing. It is installed in one of the high friction areas of footwear, in soles, ankle-foot orthosis.

- It is widely used in medical synthesis, test and many more medicines.

PET

PET is an unreinforced, semi-crystalline thermoplastic polyester derived from poly-ethylene terephthalate. Its excellent wear resistance, low coefficient of friction, high flexural modulus and superior dimensional stability make it a versatile material for designing mechanical and electro-mechanical parts. Because PET has no centerline po-rosity the possibility of fluid absorption and leakage is virtually eliminated.

Products packaged in PET include:

- Water.

- Carbonated soft drinks.

- Juice.

- Ketchup.

- Salad dressing.

- Peanut butter.

- Fresh produce.

- Baked goods.

- Frozen foods.

- Beauty & household products.

- Beer, wine, spirits and many other food and non-food items.

PET container types include:

- Bottles.

- Cups.

- Take-out containers.

Polycarbonates

Polycarbonates are condensation polymers - polyesters of phenols and carbonic acid, HO-CO-OH. They contain [-O-CO-O-] linkage.

Preparation

They are prepared by condensing Bisphenol A with diphenyl carbonate involving elimination of phenol in presence of tert. amine catalyst.

They are also prepared by condensation reaction of phosgene and sodium salt of Bisphenol A.

Properties

- Polycarbonate is a transparent thermoplastic.

- Polycarbonate has a high melting point, around 265°C.

- It has a high tensile strength and impact resistance.

- It is resistant to water and many 'organic compounds but alkalis slowly hydrolyze it.

Applications

- The great commercial success of polycarbonate is due to its unique combination of properties extreme toughness, outstanding transparency, excellent compatibility with several polymers and high heat distortion resistance.

- It finds application in making many useful articles such as safety goggles, safety shields, Telephone parts and automobile light lenses etc.

- Polycarbonates are clear thermoplastic polymers which are mainly used as molding compounds.

- CD-ROMs and baby bottles are well known examples of their use.

- Used as insulator in electronic and electrical industries.

1.4 Elastomers: Natural Rubber, Compounding and Vulcanization

Elastomers or Rubbers

Elastomers are linear polymers which have elastic properties.

Depending on the ultimate form and use, a polymer can be classified as plastic, elastomers, fiber or liquid resins. When vulcanized into the rubbery products exhibiting good strength and elongation, polymers are used as 'elastomers'. Typical examples are natural rubber, synthetic rubber, silicone rubber.

Natural Rubber

Rubber is a natural elastic polymer of isoprene. It is obtained from the milk of rubber called 'Latex'. The structure of natural rubber is as follows:

$$-(-CH_2-C=CH-CH_2-)- \\ | \\ CH_3$$

There are plenty of different rubber types available but still the largest single type used is natural rubber (NR) produced of latex from the tree Hevea Brasiliensis. The synthetic types of polymers mainly manufactured from oil have been developed either to replace or to be used together with NR or to make polymers with properties superior of NR in special areas typically with better high temperature resistance better outdoor resistance or resistance to fuels and oils.

Nowadays 90% of the Natural rubber is obtained in the form of latex from rubber trees

(Heava Brasiliensis). The latex normally contains 30-65% rubber. It can be used, as such, in the latex form or the solid rubber can be coagulated by the addition of 5% solution of acetic acid or 90% formic acid.

Ammonium or Potassium alum are also used as coagulants. Natural rubber is a highly soft and elastic material. Natural rubber is a polymerized form of isoprene (2-methyl-I, 3-butadiene),

$$H_2C=C-CH=CH_2$$
$$|$$
$$CH_3$$

Isoprene is synthesized from propylene,

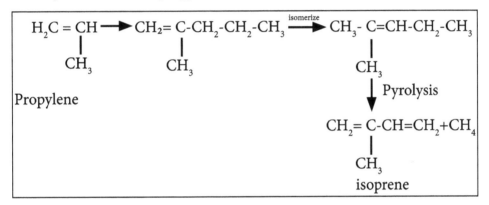

These isoprene molecules polymerize to form, long-coiled chains of cis-polyisoprene (Natural rubber).

Preparation of Natural Rubber from Latex:

- The clear liquid from the top is treated with acetic acid or formic acid to precipitate rubber.

- Latex is rubber milk containing about 30 to 45% of rubber.

- Rubber sheets are finally dried by smoking. This rubber is called 'Smoked rubber'.

- The rubber milk is diluted with water and allowed to stand for sometime.

- The precipitated rubber is collected and passed through rollers to get sheets of rubber.

- During the coagulation of rubber milk with acetic or formic acid, retardants like sodium bisulphite ($NaHSO_3$) are added to prevent oxidation of rubber. This is called 'Creep rubber'.

Different Types of Rubber

- General purpose elastomers.
- Special purpose elastomers.
- Speciality elastomers.

General Purpose Elastomers Comprise

- Natural rubber (NR).
- Polyisoprene rubber (IR).
- Styrene-butadiene rubber (SBR).
- Butadiene rubber (BR).

These types have good physical properties, good processability and compatibility, are generally economical and are typical polymers used in tyres and mechanical rubber goods with demand for good abrasion resistance and tensile properties. General purpose types constitute the largest volume of polymer used.

Special Purpose Elastomers Comprise

- Ethylene-propylene rubber (EPM and EPDM).
- Butyl rubber (IIR).
- Chloroprene rubber (CR).
- Acrylonitrile-butadiene rubber or Nitrile rubber (NBR).

They have all unique properties which cannot be matched by the general purpose types and are very important for manufacturing of industrial and automotive rubber products.

The Speciality elastomers are polymers with very special properties used in automotive, aircraft, space and offshore industries. Some of these polymers are:

Chlorosulfonated Polyethylene (CSM)

- Acrylic Rubber (ACM).
- Silicone Rubber (PMQ/PV/MQ/VMQ).
- Floursilicone Rubber (FPQ).
- Fluoroelastomers (FPM/FFKM/FEPM).
- Urethane Rubber (AU/EU).

- Epichlorohydrine Rubber (CO/ECO/GECO).

Drawbacks of Natural Rubber

- Easy aging.
- Cannot resist strong acid and oil.
- Can resist plant oil.

Compounding of Rubber

Compounding is mixing of the raw rubber with other substances so as to impart the product's specific properties, suitable for a particular job.

The following materials may be incorporated depending on the service conditions of the item to be made from it.

- Softeners and plasticizers.
- Vulcanizing agents.
- Accelerators.
- Antioxidants.
- Colouring matter/pigments.
- Reinforcing fillers.

1. Softeners and Plasticizers

These are added to give the rubber greater tenacity and adhesion. Important materials are vegetable oils, waxes, stearic acid etc.

2. Vulcanizing Agents

The main substance added is sulphur. Depending upon the nature of product required, the percentage of "S" is added, it varies between 0.15 to 32%.

Many other vulcanizing agents added to rubber are sulphur monochloride, hydrogen sulphide, benzoyl chloride etc.

3. Accelerators

These materials drastically shorten the time required for vulcanization. The most usual accelerators used are 2-mercaptol, benzthiozol and zinc alkylzanthate.

4. Antioxidants

Natural rubber undergoes oxidation. For this anti oxidation materials such as complex amines, polyphosphites are added.

But, these materials get darken in the presence of light and therefore, a substance like β-naphthol is used.

5. Colouring Matter/pigment

These are added to give the rubber product the desired colour.

For white products, titanium dioxide pigments are used while for coloured products pigments like chromium oxide-Green, ferric oxide-Red, antimony sulphide-Crimson, lead chromate-Yellow are added.

6. Reinforcing Fillers

These are added to give the strength and rigidity to the rubber products.Most common of all reinforcing fillers especially for the manufacture of motor car tyres is carbon black.

Other materials such as ZnO, $CaCO_3$, $MgCO_3$ are very widely used.

Vulcanization of Rubber

Vulcanization is achieved by heating rubber with sulphur at 1400C in CO_2 atmosphere. Vulcanization is compounding of rubber with sulphur.

Sulphur adds to the double bonds present in rubber to provide cross links between the polymer chains.2 to 4% Sulphur addition gives soft elastic rubber. When sulphur content is more than 30%, we get hard rubber called 'Ebonite'.

$$
\begin{array}{ccc}
& CH_3 & CH_3 \\
& | & | \\
-CH_2-C{=}CH-CH_2- & \xrightarrow[CO_2 \text{ Atm}]{\text{Sulphur at } 140°C} & -CH_2-C{-}{-}CH-CH_2- \\
& & \quad\quad | \quad | \\
& & \quad\quad S \quad S \\
-CH_2-C{=}CH-CH_2- & & \quad\quad | \quad | \\
& | & -CH_2-C{-}{-}CH-CH_2 \\
& CH_3 & \quad | \\
& & CH_3
\end{array}
$$

Vulcanization refers to a specific curing process of rubber involving high heat and the addition of sulfur or other equivalent curatives. It is a chemical process in which polymer molecules are linked to other polymer molecules by atomic bridges composed of sulfur atoms or carbon to carbon bonds. The end result is that the springy rubber molecules become cross-linked to a greater or lesser extent.

This makes the bulk material harder much more durable and also more resistant to chemical attack. It also makes the surface of the material smoother and prevents it from sticking to metal or plastic chemical catalysts.

This heavily cross-linked polymer has strong covalent bonds with strong forces between the chains and is therefore an insoluble and infusible thermosetting polymer. The vulcanization process is a progressive reaction and is therefore allowed for a specified time. The process is named after vulcan, Roman god of fire.

A vast array of products are made with vulcanized rubber including ice hockey pucks, tires, shoe soles, hoses and many more. Hard vulcanized rubber is known as ebonite or vulcanite and is used to make bowling balls and clarinet mouth pieces.

Reason for Vulcanizing

Uncured natural rubber is sticky can easily deform when warm and is brittle when cold. In this state it cannot be used to make articles with a good level of elasticity. The reason for inelastic deformation of unvulcanized rubber can be found in its chemical nature: rubber is made of long polymer chains. These polymer chains can move independently relative to each other which results in a change of shape.

By the process of vulcanization cross links are formed between the polymer chains so the chains can no longer move independently. As a result when stress is applied the vulcanized rubber will deform but upon release of the stress the rubber article will go back to its original shape.

Properties

- The process of vulcanizing rubber is fairly simple.

- Liquid rubber has sulfur added to it and is then heated at a high temperature.

- Elasticity Vulcanized rubber is elastic as it was before the vulcanization process assuming not more than 10% sulfur was added by weight of the mixture.

- This adds to the durability of vulcanized rubber allowing it to bend and then move back to its original shape.

Strength when rubber is vulcanized it becomes cross-linked in its chemical structure at the atomic level. This linking of stronger bonds makes vulcanized rubber over 10 times stronger than natural rubber would be. This is one of vulcanized rubber's greatest strengths as it allows rubber to be used in making more heavy duty products since it can stand up to more punishment.

Rigidity while vulcanized rubber is elastic meaning it will return to its original shape it is also 10 times more rigid than normal rubber as a result of the vulcanization process.

Uses of Vulcanized Rubber

- Vulcanized natural rubber is used for making: Rubber bands, football bladders, gloves and rubber tubes.

- 5% sulphur is used for making tyre rubber.

- Tyres and tubes of automobiles and conveyor belts for industrial use after hardening.

1.5   Synthetic Rubbers: Buna S, Buna N, Thiokol and Polyurethanes, and Applications of Elastomers

Synthetic Rubber

Synthetic rubber is created from petroleum and is classified as an artificial elastomer. This means that it is able to be deformed without sustaining damage and can return to its original shape after being stretched. Man-made rubber has many advantages over natural rubber and is used in many applications due to its superior performance. The use of synthetic rubber is much more prominent than natural rubber in most industrialized nations.

Synthetic rubber is used in a wide variety of applications. In addition to its importance in car tyres, artificial rubber is also commonly used to produce medical equipment, molded parts and belts for machinery. Many industrial hoses and seals are also created using man-made rubber.

There are several different popular varieties of synthetic rubber. These are usually created by combining chemicals in different quantities during the rubber production process. Styrene butadiene rubber (SBR) is very common and is able to withstand temperatures between -40 to 212° F (-40 to 100° C). This type of rubber is widely used in tire treads for aircraft and automobiles and also for conveyor belts and other industrial products.

1. Buna-S or Styrene Butadiene Rubber (SBR)

Buna-S is obtained by co-polymerization of butadiene and styrene in the presence of sodium catalyst. It is also called as Styrene rubber or GRS rubber.

It can be vulcanized like natural rubber. It is mainly used in manufacture of tyres. It is used as electrical insulator. It is used in making floor tiles, gaskets and footwear components, etc.

n CH_2=CH-CH=CH_2+ n C_6H_5-CH=CH_2 $\xrightarrow{\text{Cuemene bydroperoxide}}$

Butadiene   Styrene

(75% by mass)   (25% by mass)

$$\left[CH_2\text{-}CH=CH\text{-}CH_2\text{-}CH\text{-}CH_2 \right]_n$$

SBR or Buna-S or GR-S

SBR resembles natural rubber in processing characteristics as well as quality of finished products. It has less tensile strength than natural rubber.

It is used in the manufacture of automobile tyres-rubber soles, belts, hoses etc.

2. Buna-N or Nitrile Rubber

n CH_2=CH–CH=CH_2 + nCH_2 =CH-CN $\xrightarrow{\text{Copolymerization}}$

Butadiene   Acrylonitrile

$$\left(CH_2\text{-}CH=CH\text{-}CH_2\text{-}CH\text{-}CH_2 \right)_n$$
$$CN$$

Nitrile Rubber or NBR

It has low swelling, low solubility, good tensile strength and abrasion resistance even after immersion in oils. These rubbers have good heat resistance.

Nitrile rubbers are used in fuel tanks, creamery equipments, gasoline hoses etc. They are also find uses in adhesive and in the form of latex, for impregnating paper, leather and textiles.

3. Thiokols or Polysulphide rubber

GR-P is prepared by the reaction between sodium polysulphide (Na2S) and ethylene dichloride $\left(CH_2CI - CH_2CI \right)$.

$$CI\text{-}CH_2\text{-}CH_2\text{-}CI \ + \ Na_2S_x \ + \ Cl.CH_2\text{-}CH_2\text{-}CI \longrightarrow$$
$$-CH_2\text{-}CH_2\text{-}S\text{-}S\text{-}CH_2\text{-}CH_2\text{-}$$
$$\downarrow \downarrow$$
$$S \ S$$

Thiokol rubber

Thiokol rubber possesses extremely good resistance to mineral oils, fuels, solvents, oxygen, ozone and sunlight. It is also impermeable to gases. It cannot be vulcanized. Thiokol rubber are used for barrage balloons, life rafts and jackets which are inflated by CO_2. Polysulphide rubbers are also used for lining hoses for conveying gasoline and oil and for printing rolls.

It does not have properties of rubber. It can withstand cold but not heat.

4. Polyurethanes or Isocynate Rubber

Polyurethane rubbers are highly resistant to oxidation, many organic solvents. These are attacked by acids and alkalies. Polyurethane foams are light, tough and heat resistant, abrasion resistant, weathering resistant and chemical resistant.

These rubbers are mainly used in making surface coatings and manufacture of foams and spandex fibres.

Applications of Elastomers

- Construction: sealants, structural glazing.

- Moldmaking: prototyping, restoration, furniture, ornaments.

- Household: gaskets, o-rings.

- Automotive: cure-in-place gaskets, vibration dampers.

- Electronics: adhesives.

Industrial Applications of Elastomeric Materials

1. Gaskets

Gaskets are mechanical seals that fill the space between imperfect mating surfaces. Gaskets are utilized to tightly fill the space it was designed for which would include any minor irregularities. Gaskets provide insulation and prevent leakage of gases or liquids between joined surfaces under compressive force.

Most gaskets are customized in size and shape in accordance with the profiles of the joined surfaces. The actual elastomer type/grade to be employed will depend on a variety of factors such as service temperature, mechanical properties within the application and environmental properties ranging from weathering/ozone/UV considerations to chemical resistance concerns (i.e. acids, salt water, oils, hydraulic fluids, etc.)

There are many different types of gasket applications. Perhaps one of the most recognizable is found in the automotive industry; the head gasket. Head gaskets in automotive engines can be made from steel, copper, composite or elastomeric material.

This gasket resides between the engine block and the cylinder heads in an internal combustion engine. This gasket seals the cylinders to provide optimal compression and prevent leakage of combustible air/fuel mix, coolant and/or motor oil.

2. Seals & O-Rings

Similar to gasket applications, elastomers are heavily utilized in seal applications including hydraulic, pneumatic and O-Ring deployments. Most industrial machinery will employ the use of one of these types of seals.

Hydraulic seal applications provide an interesting challenge for elastomeric materials because their function goes beyond just the prevention of fluid leakage. These seals must also withstand high pressures, extreme temperature ranges and transverse forces within the cylinder. The utilization of hydraulic seals enables fluid power to be converted to linear motion.

Pneumatic seals are typically used in lower pressure environments where rotary or reciprocating motions take place within a cylinder or valve. They can be exposed to high operating speeds where the pressure is not high and require minimal lubrication. Some examples of pneumatic seals are piston seals, rod seals and flange packings.

The O-Ring is a loop of elastomer with a rounded cross-section that typically fits within a groove and gets compressed between the assemblies of two or more parts. O-Rings can be utilized in static or dynamic applications and tend to be popular because they are generally cost effective, easy to manufacture, simple to assemble and reliable.

3. Noise Reduction and Dampening

The properties of elastomers make them ideal for many noise reduction and dampening functions. Many elastomeric configurations can reduce resonant vibration and resulting airborne sound. They yield effective vibration damping over a broad temperature and frequency range.

Large scale industrial fans also use elastomers to dampen sound. The utilization of elastomeric fan mounts prevents the transfer of structural vibration from the fan to its mounting structure. Reducing this vibration means less sound radiation and quieter operation.

Examples and Applications of Elastomer Plastic Materials

- Natural rubber - material used in the manufacture of gaskets, shoe heels.

- Polyurethanes - Polyurethanes are used in the textile industry for the manufacture of elastic clothing such as lycra, also used as foam, wheels, etc.

- Polybutadiene - elastomer material used on the wheels or tires of vehicles, given the extraordinary wear resistance.

- Neoprene - Material used primarily in the manufacture of wetsuits is also used as wire insulation, industrial belts, etc.

- Silicone - Material used in a wide range of materials and areas due their excellent thermal and chemical resistance, silicones are used in the manufacture of pacifiers, medical prostheses, lubricants, mold, etc.

1.6 Composite Materials and Fiber Reinforced Plastics, Biodegradable Polymers and Conducting Polymers

Composite Materials

A composite material (also called a composition material or shortened to composite which is the common name) is a material made from two or more constituent materials with significantly different physical or chemical properties that, when combined, produce a material with characteristics different from the individual components.

The individual components remain separate and distinct within the finished structure. The new material may be preferred for many reasons: common examples include materials which are stronger, lighter or less expensive when compared to traditional materials.

Fiber Reinforced Plastics

Fiber Reinforced Plastics (FRP) is the generic term for a uniquely versatile family of composites used in everything from chemical plant to luxury power boats.

Fiber reinforced plastic (FRP), is also known as a fiber reinforced polymer, it is in fact a fiber reinforced plastics composite material constituting a polymer matrix blended with certain reinforcing materials, such as fibers. The fibers are generally basalt, carbon, glass or aramid, in certain cases asbestos, wood or paper can also be used.

FRP Provides Following Properties

- Light weight.

- Good electrical insulating properties.

- High strength-to-weight ratio (kilo-for-kilo it's stronger than steel).

- High levels of stiffness.

- Design freedom.

- Chemical resistance.

- Retention of dimensional stability across a wide range of temperatures.

Applications

- Building and Construction.

- Transportation.

- Marine industry.

- Chemical plant and pipes.

Biodegradable Polymers

Polymers are the synthetic and natural macromolecules composed of smaller units called monomers. Many synthetic polymers are produced and utilized because they can resist chemical and physical degradation.

These polymers which are resistant to degradation results in disposal problems when their usefulness ceases. Research has shown that substitution of natural monomers into synthetic polymers produces polymers that are more easily biodegraded.

A biodegradable polymer is a polymer in which the degradation results from the action of naturally occurring microorganisms such as bacteria, fungi or algae. These biodegradable polymers are largely used in medical application where they undergo degradation by chemical hydrolysis.

Biodegradable polymers are a specific type of polymer that breaks down after its intended purpose to result in natural byproducts such as gases (CO_2, N_2), water, biomass and inorganic salts.

Polymers prepared from glycolic acid and lactic acid has found a multitude of uses in the medical industry, beginning with the biodegradable sutures first approved in 1960s. Since that time, diverse products based on lactic and glycolic acid and on other materials, including poly(trimethylene carbonate), poly(dioxanone) copolymers and poly(caprolactone) homopolymers and copolymers have been accepted for use as medical devices.

In addition to these approved devices, a great deal of research continues on polyanhydrides, polyorthoesters, polyphosphazenes and other biodegradable polymers.

Some examples of the biodegradable polymers are shown below:

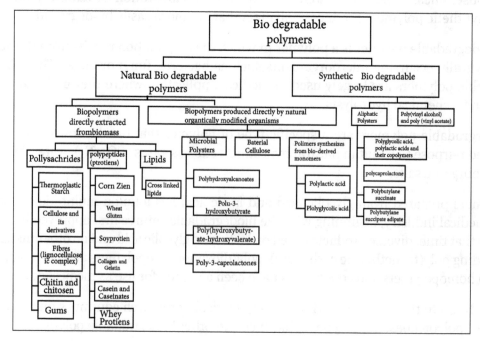

poly(DTE) carbonate

Bis(p-carboxyphenoxy)propane (PCPP)
Poly(PCPP-SA anhydride)   Sebacic acid(SA)

DTE= desaminotyrosyltyrosine
ethyl ester
                                    Poly(lactic acid-co-glycolic acid)

Biodegradable polymer.

Classification of Biodegradable Polymers

Biodegradable polymers can be divided into three broad classifications:

- Natural polymers.

- Synthetic polymers.

- Modified Natural polymers.

Advantages of Biodegradable Polymers

- Localized delivery of drug.
- Sustained delivery of drug.
- Stabilization of drug.
- Decrease in dosing frequency.
- Reduce side effects.
- Improved patient compliance.
- Controllable degradation rate.

Conducting Polymers

Conducting polymers (CP) are long chain having current flowing properties. To make the polymer materials conductive they are doped with atoms that donate negative or positive charges (oxidizing or reducing agents) to each unit, enabling current to travel down the chain. Depending on the dopant, conductive polymers exhibit either p-type or n-type conductivity.

Conducting polymers can be classified into the following types:

```
                    Conducting polymers
                           |
        +------------------+------------------+
        ↓                                     ↓
   Intrinsically                    Extrinsically conducting
conducting polymers                        polymers
        |                                     |
   +----+----+                      +---------+---------+
   ↓         ↓                      ↓                   ↓
C.P. having  Doped conducting   Conducting          Blended C.P.
conjugation  polymers      element filled polymers
```

Intrinsically Conducting Polymers

1. C.P. having Conjugated π–Electrons in the Backbone

Such polymers contains conjugated π–electron in the backbone which increases their conductivity to a large extent. This is because overlapping of conjugated π–electrons over the entire backbone results in the form of valence bonds as well as conduction bonds, which extends over the entire polymer molecule.

For example-Polypyrrole: It is obtained by electro-polymerization of pyrrole.

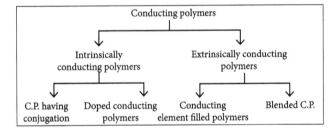

Polypyrrole is an inherently conducting polymer due to inter chain dopping of electrons. It can be easily prepared by oxidative polymerization of the monomer pyrrole.

2. Doped Conducting Polymers

It is of two types:

• p–doping (oxidative doping).

• n–doping (reductive doping).

p–doping

It is done by oxidation process. (i.e., removal of e– from the polymer pi – back bone). This formation is known as polaron. A second oxidation of this polaron, followed by radical recombination yields two positive charge carriers on each chain which are mobile.

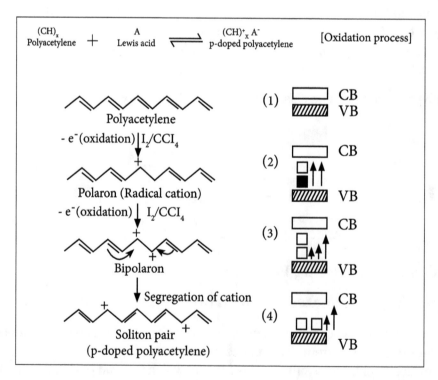

n–doping

It is done by reduction process (addition of an e– to the polymer).

$$ \underset{\text{Polyacetylene}}{(CH)x} \quad + \quad \underset{\text{Lewis acid}}{B^+} \quad \rightleftharpoons \quad \underset{\substack{\text{n-doped} \\ \text{polyacetylene}}}{(CH)^-_x \ B^+} $$

It forms polaron and bipolaron in two steps. This followed by recombination of radicals yields two negative charge carriers on each chain of polyacetylene.

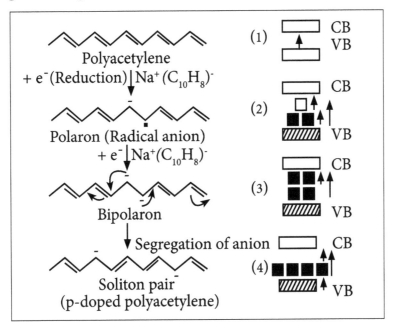

Extrinsically Conducting Polymers

It is of two types:

- Conductive Element filled polymers.

- Blended conducting polymers.

1. Conductive Element Filled Polymer

In this, the polymers act as the binder to hold the conducting element (such as carbon black, metallic fibers, metallic oxides etc.) together in the solid entity.

Minimum concentration of conductive filler which should be added so that polymer starts conducting is known as percolation threshold. Carbon black is used as filler which has very high surface area ($1000m^2/gm$) more porosity and more of a filamentous properties. It bears good conductive properties and low in cost, light in weight, as well as durable.

It is used in hospital operating theatres where it is essential that static charges did not build up leading to explosion involving on aesthetics.

2. Blended Conducting Polymer

It is obtained by blending a conventional polymer with a conducting polymer either by physical or chemical change. Such polymers can be easily processed and possess better physical, chemical and mechanical properties. The important commercially produced conducting polymers include:

- Polyacetylene polymers, e.g., poly-p-phenylene, polyquinoline, polyphenylene-co-vinylene, poly-m-phyenylene sulphide, etc.

- With condensed aromatic rings, e.g., polynanthrylene, polyphenanthrylene, etc.

- With aromatic heteroaromatic and conjugated aliphatic units, e.g., polypyrrole, polythiophene, polyazomethane, polybutadienylene, etc.

2

Fuel Technology

2.1 Fuels: Classification, Calorific Value, HCV, LCV, Dulong's Formula and Bomb Calorimeter

Fuels

Fuels are materials such as coal, gas or oil that is burned to produce heat or power. The various types of fuels like solid, liquid and gaseous fuels are available for firing in boilers, furnaces and other combustion equipments. The selection for right type of fuel depends on various factors such as availability, storage, handling, landed cost of fuel and pollution.

The knowledge of the fuel properties helps in selecting the right fuel for the right purpose and efficient use of the fuel.

Classification

Fuels may broadly be classified in two ways, i.e. (a) according to the physical state in which they exist in nature as solid, liquid and gaseous and (b) according to the mode of their procurement as natural and manufactured. Some of the fuels are:

- Class 1: Solid combustible materials that are not metals. Ex: wood, paper, cloth, trash, plastics.

- Class 2: Any non-metal in a liquid state on fire. Ex: flammable liquids: Gasoline, oil, grease, acetone.

- Class 3: Electrical Energized electrical equipment. As long as it's "plugged in," it would be considered a class C fire.

- Class 4: Metals: Potassium, sodium, aluminum, magnesium.

Calorific Value

Calorific Value (CV) is a measure of heating power and it is dependent upon the composition of the gas. The CV refers to the amount of energy released when a known volume of gas is completely combusted under a specified conditions.

The calorific value is the measurement of heat or energy produced and it is measured either as gross calorific value or net calorific value. The difference in the latent heat of condensation of the water vapour is produced during the combustion process. Gross calorific value (GCV) assumes all the vapour which is produced during the combustion process is fully condensed.

Net calorific value (NCV) assumes that the water leaves with the combustion products without it is fully being condensed. Fuels should be compared based on the net calorific value.

Thus, the calorific value of coal varies considerably depending on the ash, moisture content and the type of coal while calorific value of fuel oils are much more consistent.

Higher Calorific Value (HCV)

Higher calorific value of a fuel portion is defined as the amount of heat evolved when a unit weight or the volume in the case of gaseous fuels is completely burnt and the products of the combustion is cooled to the normal conditions (with water vapor condensed as a result). The heat contained in the water vapor should be recovered in the condensation process.

Lower Calorific Value (LCV)

Lower calorific value of a fuel portion is defined as the amount of heat evolved during a unit weight or the volume in the case of gaseous fuels is completely burnt and the water vapor leaves with the combustion products without being condensed.

Some typical values	H.C.V.	L.C.V.
Anthracite coal	34.583	33.913 MJ/kg
Wood	15.826	14.319
Petrol	46.892	43.710
Diesel	45.971	43.166
U.K. Natural gas	36.38	32.75 MJ/m³ **
Hydrogen	11.92	10.05

** at SSL conditions (15 °C & 101.325 kPa]

Principle

The gross calorific value of a solid or liquid fuel when burnt in excess air or oxygen is calculated. The heat liberated is absorbed into surrounding water and copper calorimeter.

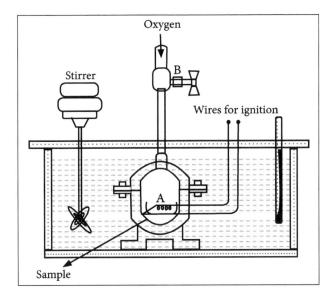

Construction: The bomb calorimeter consists of a cylindrical steel vessel (bomb A) with an airtight screw lid and an inlet valve B, for pumping oxygen. The bomb has a platinum crucible with a loop of wires or electrical ignition coil for initial combustion of fuel. The bomb is kept in a rectangular copper vessel (calorimeter) containing weighed amount of water.

The equipment is provided with a mechanical stirrer for uniform heat distribution and a Beckmann thermometer to note the temperature. Calorimeter is enclosed in a jacket to minimize the heat exchange with surroundings.

Working: A known mass of the fuel is made into a pellet and taken in the crucible and oxygen is passed through the inlet valve. A known mass of water is taken in the calorimeter and is closed with the lid. The initial temperature of water is noted. The fuel is ignited by passing current through the coil. The heat released due to burning of fuel is absorbed by water. The temperature of water rises. The final temperature of water is noted on the thermometer.

Calculation

Let,

Mass of fuel = m kg

Mass of water = w_1 kg

Water equivalent of calorimeter = w_2 kg

Initial temperature of water = t_1 °C

Final temperature of water = t_2 °C

Specific heat of water $(s) = kJ\ kg^{-1}\ °C^{-1}$

$$\therefore GCV\,(\text{Solid fuel}) = \frac{(w_1 + w_2) \times (t_2 - t_1) \times S}{m}\,kJ\ kg^{-1}$$

NCV can be calculated if the % of H_2 in the fuel is known.

$$2\,H + \tfrac{1}{2}\,O_2 \rightarrow H_2O$$

If the fuel contains x% hydrogen, NCV of the fuel is calculated as follows:

2 atoms of hydrogen produces one molecule of water.

2g of hydrogen produces 18 g of water.

i.e., x g of hydrogen produces 9 x g of water.

x % hydrogen (9/100) x g of water = 0.09 x g of water.

NCV = GCV - latent heat of steam formed

= GCV - 0.09 × % of H_2 × latent heat of steam

= GCV - 0.09 × % of H_2 × 587 cal/g (Or)

NCV = GCV - 0.09 × % of H_2 × 2454 kJ kg-1

Dulong's Formula

Dulong suggested a formula for the calculation of the calorific value of the solid or liquid fuels from their chemical composition which is as given below.

Gross calorific value,

Or

$$H.H.V. = \frac{1}{100}\left[33800C + 144000\left(H - \frac{O}{8}\right) + 9270\,S\right] KJ/kg$$

Where C, H, 0 and S are carbon, hydrogen, oxygen and sulphur in percentages respectively in 100 kg of fuel. In the above formula the oxygen is assumed to be in combination with hydrogen and only extra surplus hydrogen supplies the necessary heat.

Bomb Calorimeter

The calorific value of solid and liquid fuels is determined in the laboratory by 'Bomb calorimeter'. It is so named because its shape resembles that of a bomb. Figure shows the schematic sketch of a bomb calorimeter.

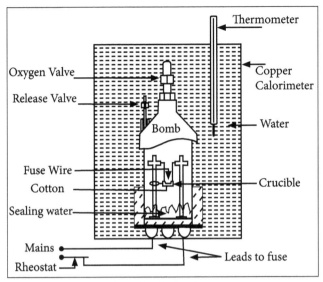

Bomb calorimeter.

The calorimeter is made of a austenitic steel which provides considerable resistance to corrosion and enables it to withstand high pressure. In calorimeter there is a strong cylindrical bomb in which combustion occurs. The bomb has two valves at the top. One supplies oxygen to bomb and other releases the exhaust gases.

A crucible in which a weighted quantity of fuel sample is burnt is arranged between two electrodes as shown in figure. The calorimeter is fitted with water jacket. which surrounds bomb. To reduce the losses due to radiation, calorimeter is further provided with a jacket of water and air. A stirrer for keeping the temperature of water uniform and a thermometer to measure the temperature up to an accuracy of 0.001°C are fitted through the lid of the calorimeter.

Procedure

To start with, about 1 gm of fuel sample is accurately weighed into crucible and a fuse wire (whose weight is known) is stretched between the electrodes. It should be ensured that wire is in close contact with the fuel. To absorb the combustion products of the sulphur and the nitrogen 2 ml of water is poured in the bomb. Bomb is then supplied with pure oxygen through the valve to an amount of 25 atmosphere.

Bomb is then placed in the weighed quantity of water, in the calorimeter. The stirring is started after making necessary electrical connections and when the thermometer indicates a steady temperature fuel is fired and the temperature readings are recorded after 1/2 minute intervals until maximum temperature is attained. The bomb is then removed, pressure is slowly released through the exhaust valve and the contents of the bomb are carefully weighed for further analysis.

The heat released by the fuel on combustion is absorbed by the surrounding water and the calorimeter. From the above data the calorific value of the fuel can be found in the following way:

Let,

w_f = Weight of fuel sample (kg).

w = Weight of water (kg).

C = Calorific value (higher) of the fuel (kJ/kg).

w_e = Water equivalent of calorimeter (kg).

t_1 = Initial temperature of water and calorimeter.

t_2 = Final temperature of water and calorimeter.

t_c = Radiation corrections.

c = Specific heat of water.

Heat released by the fuel sample = $w_f \times C$

Heat received by water and calorimeter = $(w_w + w_e) \times c \times \left[(t_2 - t_1) + t_c\right]$.

Heat lost = Heat gained

$$\therefore w_f \times C = (w + w_e) \times c \times \left[(t_2 - t_1) + t_c\right]$$

$$C = \frac{(w + w_e) \times c \times \left[(t_2 - t_1) + t_c\right]}{w_f}$$

[Value of c is 4.18 in SI units and unity in MKS units.]

Problems

1. Let us calculate the gross calorific value and net calorific value of a sample of coal 0.5g which when burnt in a bomb calorimeter, raised the temperature of 1000g of water from 293K to 296.4K. The water equivalent of calorimeter is 350g. The specific heat of water is 4.187 kJ $kg^{-1}K^{-1}$, latent heat of steam is 2457.2 kJ kg^{-1}. The coal sample contains 93% carbon, 5% hydrogen and 2% ash.

Solution:

Given:

- m = Mass of the fuel = 0.5 g.

- w_1 = Mass of water taken = 1000 g.

- w_2 = Water equivalent of calorimeter = 350 g.

- t_1 = Initial temperature of water = 293 K.

- t_2 = Final temperature of water = 296.4 K.

- S = Specific heat of water = 4.187 kJ kg^{-1} K.

- $\text{GCV}(\text{solid fuel}) = \dfrac{(w_1 + w_2) \times (t_2 - t_1) \times S}{m}$ kJ kg^{-1}.

Formula to be used:

NCV (solid fuel) = GCV - latent heat of steam formed.

= GCV - (0.09 ×% of H$_2$) × latent heat of steam

$$\text{GCV}(\text{solid fuel}) = \frac{(w_1 + w_2) \times (t_2 - t_1) \times s}{m} \text{ kJ kg}^{-1}$$

$$= \frac{(1000 + 350)\,\text{g} \times (296.4 - 293)\,\text{K} \times 4.187\,\text{kJ kg}^{-1}\text{K}^{-1}}{0.5\,\text{g}}$$

$$= \frac{1350\,\text{g} \times 3.4\,\text{K} \times 4.187\,\text{kJ kg}^{-1}\text{K}^{-1}}{0.5\,\text{g}}$$

$$= 38,437 \text{ kJ kg}^{-1}$$

NCV (solid fuel) = GCV - latent heat of steam formed.

= GCV - (0.09 ×% of H$_2$) × latent heat of steam

= 38437 kJ kg^{-1} – (0.09 × 5) × 2457.2 kJ kg^{-1}

= 38437 kJ kg^{-1}– 1106 kJ kg^{-1}

= 37,331 kJ kg^{-1}

2. Let us calculate the gross and net calorific value of a coal sample from the following data obtained from bomb calorimeter experiment.

Solution:

Given:

- Weight of coal (m) = 0.73 g.

- Weight of water taken in calorimeter $(w_1) = 1500$ g.

- Water equivalent of calorimeter $(w_2) = 470$ g.

- Initial temperature $(t_1) = 25°C$.

- Final temperature $\left(t_2\right) = 27.3°C$.

- Percentage of Hydrogen in coal sample = 2.5%.

- Latent heat of steam = 587 Cal/g.

$$GCV\left(\text{solid fuel}\right) = \frac{\left(w_1 + w_2\right) \times \left(t_2 - t_1\right) \times S}{m} \text{ kJ kg}^{-1}$$

Formula to be used:

NCV (solid fuel) = GCV - latent heat of steam formed.

$$= GCV - \left(0.09 \times \% \text{ of } H_2\right) \times \text{latent heat of steam}$$

$$GCV\left(\text{solid fuel}\right) = \frac{\left(w_1 + w_2\right) \times \left(t_2 - t_1\right) \times s}{m} \text{kJ kg}^{-1}$$

$$= \frac{\left(1500 + 470\right) \times 10^3 \text{ kg} \times \left(27.3 - 25\right)°C \times 4.187 \text{ kJ kg}^{-1} °C^{-1}}{0.73 \times 10^3 \text{ kg}}$$

$$= 25988 \text{ kJ kg}^{-1}$$

Net calorific value = HCV - Heat released by the condensation of steam

$$= GCV - 0.09 \times \% H_2 \times \text{Latent heat of steam}$$

$$= 25988 - 0.09 \times 2.5 \times 2454$$

$$= 25435.85 \text{ kJ kg}^{-1}$$

2.2 Coal: Proximate, Ultimate Analysis and Significance of the Analyses

Coal

Coal is a natural fuel formed by the slow carbonization of vegetable matter buried under the earth some thousands of years ago. It is classified into four kinds based on the carbon content and the calorific value. They are:

- Peat.

- Lignite.

- Bituminous Coal.

- Anthracite Coal.

1. Peat

It is the first stage of formation of coal from wood. Peat is brown in color and fibrous jelly-like mass. It contains 80-90% moisture. The calorific value of peat is 5400kcal/kg. The composition of peat is C=57%, H=6%, O=35%, Ash=2.5%. It is a low grade fuel due to high water content. It is used as a fertilizer and packing material.

2. Lignite

Lignite is immature form of coal. It contains 20-60% moisture. The air dried Lignite contains C=60-70%, O=20%. This burns with a long smoky flame. The calorific value of lignite is 6500-7100 kcal/kg.

Uses

- This is used as a domestic fuel.

- It is used as a boiler fuel for steam raising.

- It is used in the manufacture of producer gas.

3. Bituminous Coal

It is a high quality fuel. Its moisture content is 4%. Its calorific value is 8500 kcal/kg. Its composition is C=83%, O=10%, H=5% and N=2%.

Uses

- This is used in metallurgy.

- It is used for making coal gas.

- This is used in steam raising.

- It is also used for domestic heating.

4. Anthracite Coal

It is the superior form of coal. it contains C=92-98%, O=3%, H=3% and N=0.7%. It burns without smoke. It's calorific value is 8700 kcal/kg.

Uses

- This is used for steam raising and house hold purposes.

- It is used for direct burning in boilers and in metallurgy.

- It is used in coal tar distillation.
- It is used in thermal power plant.
- It is used in glass furnaces.

Proximate and Ultimate Analysis

Proximate Analysis

Proximate analysis includes the determination of the moisture, volatile, ash and the fixed carbon content.

- Determination of Moisture content: 1 g of finely powdered air dried sample is taken in a crucible and it is heated in an electrically heated hot air oven at 105°C-110°C for 1 hr. After heating, the crucible is taken out side from the air oven and cooled in a desiccator and weighed.

$$\text{Percentage of moisture content} = \frac{\text{Loss in weight of coal}}{\text{Weight of coal taken}} \times 100$$

- Determination of Volatile matter: The dried sample which is left in the crucible along with the lid that is heated in a muffle furnace at a temperature of 950 ± 25°C for 7 minutes and then cooled in a desiccator and weighed.

$$\text{Percentage of volatile content} = \frac{\text{Loss in weight of coal}}{\text{Weight of coal taken}} \times 100$$

- Determination of Ash content: The residual sample is obtained after the two above experiments in the crucible which is being heated in the furnace at 750°C for 30 minutes without the lid. Then it is cooled in a desiccator and weighed.

$$\text{Percentage of ash content} = \frac{\text{Weight of ash left}}{\text{Weight of coal taken}} \times 100$$

- Determination of Fixed carbon: Here, the fixed carbon content can be determined indirectly by subtracting the percentage of moisture, volatile and ash content from 100 % of fixed carbon = 100% of (moisture + volatile matter + ash).

Significance

- If it has higher percentage of moisture content then it lowers the calorific value of coal. Hence lower the moisture content better the quality of coal.
- If it has higher percentage of volatile matter then it reduces the calorific value of coal. Low volatile matter shall also reduce the coking property of coal.
- The ash being non-combustible reduces the calorific value of coal. The ash

deposition also causes problems in the furnace walls and the ultimate disposal of ash is also a problem.

- If it has higher percentage of fixed carbon, then higher is the calorific value and better is the quality of coal.

Ultimate Analysis

It refers to determination of weight percentage of carbon, hydrogen, nitrogen, oxygen and sulfur.

1. Carbon and Hydrogen: A required amount of coal sample is taken and burnt in a current of O_2 in combustion apparatus whereby CO_2 and H_2O are formed. The CO_2 and H_2O are absorbed by previously weighed tubes containing KOH and anhydrous $CaCl_2$. The increase in weight gives the C and H content as follows:

$$C + O_2 \rightarrow CO_2$$

$$H_2 + 1/2\,O_2 \rightarrow H_2O$$

$$2KOH + CO_2 \rightarrow K_2CO_3 + H_2O$$

$$CaCl_2 + 7H_2O \rightarrow CaCl_2 \cdot 7H_2O$$

Increase in weight of KOH tube % of C= × 12/44 × 100.

Weight of coal taken.

Increase in wight of $CaCl_2$ tube % of H = × 2/18 × 100.

2. Nitrogen: About 1 g of accurately weighed coal sample is taken in a long necked flask along with the conc. H_2SO_4, K_2SO_4 and heated. Then it is treated with excess of NaOH and the liberated NH_3 is absorbed in standard acid solution.

The excess acid is back-titrated with standard NaOH solution. From the volume of acid consumed, the N content is calculated as follows:

$$\% \text{ of N} = \frac{\text{Volume of acid consumed} \times \text{Normality}}{\text{Weight of coal taken}} \times 1.4$$

3. Sulphur: While determining the calorific value of a coal sample in a bomb calorimeter, the S in the coal is converted to sulphate. Finally, the washings containing sulphate is treated with dilute HCl and $BaCl_2$ solution which precipitates $BaSO_4$ which is filtered in a sintered glass crucible, washed with water and heated to a constant weight.

$$\% \text{ of S} = \frac{\text{Weight of } BaSO_4}{\text{Weight of coal taken}} \times \frac{32}{233} \times 100$$

4. Oxygen content = 100 – % of (C + H + S + N).

Significance

- Higher percentage of C and H increases the calorific value of coal and hence better is the coal.

- Higher the percentage of O_2 lower is the calorific value and lower is the coking power. Also O_2 when combined with hydrogen in the coal, hydrogen available for combustion becomes unavailable.

- S although contributes to calorific value is undesirable due to its polluting properties as it forms SO_2 on combustion.

Problem

The ultimate analysis of a coal sample indicates Carbon = 84%, Sulphur = 1.5%, Nitrogen = 0.6%, Hydrogen = 5.5% and Oxygen = 8.4%. Let us calculate the GCV.

Solution:

Given data:

C = 84%

S = 1.5%

N = 1.6%

H = 5.5%

O = 8.4%

$$GCV = \frac{1}{100}\left[8080C + 34500\left[H - \frac{O}{8}\right] + 2240S\right] Kcal/kg$$

$$= \frac{1}{100}\left[8080 \times 84 + 34500\left[5.5 - \frac{8.4}{8}\right] + 2240 \times 1.5\right] Kcal/kg$$

$$= 8356.05 \ K.cal/kg$$

2.3 Liquid Fuels, Petroleum, Refining, Cracking, Petrol, Diesel Knocking and Anti-knock Agents

Liquid Fuels

Liquid fuels are those which are combustible, energy-generating substances and play vital role in transportation and economy. Most widely used liquid fuels are derived from fossil fuel/petroleum/ crude oil. Some important liquid fuels are petrol, kerosene, diesel, etc.

The liquid fuels can be classified as follows:

- Natural or crude oil.

- Artificial or manufactured oils.

Advantages

- Liquid fuels occupy less storage space than solid fuels.

- As compared to solid fuels, they have a high calorific value.

- They can be easily transported through pipes.

- Liquid fuels do not yield any ash or residue during burning.

- The burning process of liquid fuels is clear.

- The combustion is uniform and very easily controllable.

- For complete combustion of liquid fuels, less air is required than that of the solid fuels and hence their thermal efficiency is high.

- They can be used in IC engines, boilers and gas turbines.

- They do not undergo spontaneous combustion.

Disadvantages

- When the liquid fuels undergo incomplete combustion, they give unpleasant odour.

- In comparison with solid fuels they are costly.

- Risk of fire hazards is more in the case of inflammable and volatile liquid fuels. Thus, they should be stored and transported more carefully.

- Some amount of liquid fuels may escape due to evaporation during storage.

- Special type of burners and sprayers are required for effective combustion.

Petroleum

Petroleum is the naturally available liquid fuel. This is a dark greenish-brown viscous oil found deep in earth's crust. It is composed of various hydrocarbons with a small amount of other organic compounds as impurities.

Refining of Petroleum

The process of purification and separation of various fractions present in petroleum by fractional distillation is called refining of petroleum. Refining is done in oil refineries.

Petroleum is a complex mixture of organic liquids (hydrocarbons) also known as crude oil or fossil fuel. It is formed from the fossilized dead plants and animals by exposure to heat and pressure in the Earth's crust.

It is a viscous dark coloured, foul-smelling liquid along with water and soil particles. Hence, it is necessary to separate these hydrocarbons into useful products and this process is known as fractional distillation. In this process, products are separated depending on boiling points, known as refining of petroleum. Refining of petroleum involves the following 3 steps.

Step 1: Separation of Water by Cottrell's Method

Petrol or crude oil is the emulsion of oil and salt water and these colloidal water droplets coalesce to form large drops which can separate out from oil when the crude oil is sent through two highly charged electrodes.

Step 2: Removal of Sulphur Compounds

Crude oil is treated with copper oxide, sulphur reacts with copper to form copper sulphide precipitate, which is removed by filtration.

Doctors Sweetening Process

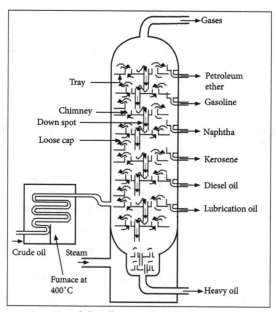

Fractional distillation of crude petroleum.

The process was described by G. Wendt and S. Diggs. Here, crude oil is treated with sodium plumbate, i.e., doctors solution, converts mercaptans in sour gasoline into disulphide.

$$\underset{\text{(sodium plubate)}}{NO_2PbO_2} + \xrightarrow{\text{powerded S in presence of NaOH}} R-S-S-R + PbS + 2NaOH \underset{\text{(alkyl disulphide)}}{}$$

Step 3: Fractional Distillation

In an iron retort, the crude oil is heated to about 400-430°C. Here, all volatile matter are evaporated, components which are not volatile like tar and asphalt are settled at the bottom of the column. The hot vapours are then passed through a distillation column, shown in Figure.

The distillation chamber is a steel cylindrical tube about 31 m height and 3 m in diameter and inside, the chamber trays are fitted at short distances. Every tray is having many holes and an up going short tube with a bubble cap.

At different heights of chamber, the vapours go up, begin to cool and condense in fractions. Fractions which are having higher boiling point condenses first and lower boiling fractions one after other.

Cracking

Cracking is a process by which the hydro carbons of high molecular mass are decomposed into hydrocarbons of low molecular mass by heating in the presence or absence of a catalyst. Generally, aluminum silicates are used as catalyst. Example:

$$\underset{\text{decane}}{C_{10}H_{22}} \xrightarrow{\text{cracking}} \underset{\text{(n-pentane + Pentene)}}{C_5H_{12} + C_5H_{10}}$$

Gas oil and Kerosene contain hydrocarbons of high molecular mass and boiling. They are unsuitable as fuel in automobiles. Hence they are decomposed into hydrocarbons of low molecular mass and low boiling point.

Fluidized (Moving) Bed Catalytic Cracking

Principle

In this the finely divided catalyst is kept agitated by cracking oil, so that it can be handled like a fluid system. One of the advantages of the FCC is that it gives very good contact between oil and catalyst. Therefore high yield of petrol is obtained.

Conditions

Feed stalk : Vapours of heavy oil fraction.

Catalyst: Al_2O_3 + SiO_2 (Alumina+ silicon dioxide)

Temperature: 530°C

Pressure: Little above the normal pressure.

Yield: Very high, usually 10 gallons per day.

The finely divided catalyst is fluidized by the upward passage of the feed stalk in the cracking chamber. Cracked vapors are continuously withdrawn from the cracking chamber and fed into fractionating column where it gets separated into gas and gasoline. The uncracked oil may be cracked in the second stage of cracking process.

The spent catalyst is continuously withdrawn from the bottom of the cracking chamber and transported into regenerators by a stream of air in which the carbon deposited on a catalyst is burnt off using hot air. The regenerated catalyst is mixed with fresh feed stalk and returned to cracking chambers.

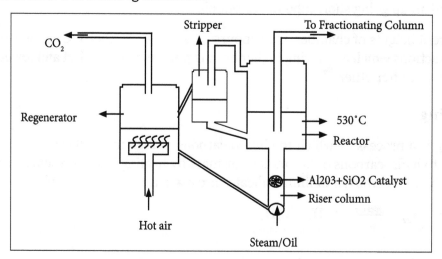

Synthetic Petrol

Manufacture of Synthetic Petrol (Bergius Process)

Coal is ground and made into a paste with heavy recycle oil and a catalyst like tin oleate. The paste is sent along with H2 at 250-350 atm pressure into a converter which is maintained at 450°C - 500°C temperature.

The un-reacted coal is filtered-off and the liquid product distilled. Hydrogen combines with coal to form saturated hydrocarbons which decompose at high temperature yielding low-boiling hydrocarbons. The crude oil is fractionated to get (i) gasoline (ii) middle oil and (iii) heavy oil which is recycled. Middle oil is hydrogenated in vapour phase with catalyst to yield more gasoline. Yield of gasoline is 60% of the coal dust.

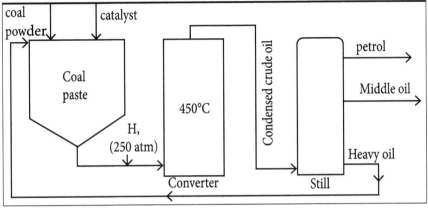

Bergius process.

Synthesis of Petrol by Fischer-Tropsch Process: (Indirect Conversion of Coal)

This method involves the following steps:

1. Production of water gas: Water gas $(CO+H_2)$ is obtained by passing steam over white hot coal.

$$\underset{\text{Coke\; steam}}{C+H_2O} \xrightarrow{1200°C} \underset{\text{Water gas}}{CO+H_2}$$

2. The water gas is mixed with hydrogen and the mixture is purified by passing through Fe2O3 and then passed into a mixture of $Fe_2O_3+Na_2CO_3$.

   Fe_2O_3 is used to remove H_2S.

   $Fe_2O_3 + Na_2CO_3$ is used to remove organic sulphur compounds.

Production of synthesis gas,

Water gas obtained above is freed from dust, H_2S and organic Sulfur compounds and blend water gas with hydrogen to form synthesis gas $(CO+2H_2)$.

3. Hydrogenation of carbon monoxide: the Synthesis gas $(CO+2H_2)$ is compressed to 5-10 atm pressure and admitted into a catalytic reactor containing the catalyst (mixture of cobalt (100 parts), thoria (5 parts) and magnesia (8 parts)).

The reactor is heated to about 250°C. Hydrogenation, reactions takes place to form saturated and unsaturated hydrocarbon. These mixture of saturated and unsaturated hydrocarbons are passed through a fractionating column for separation of petroleum fractions. It is produced as a result of polymerization.

$$n\; CO+2n\; H_2 \rightarrow C_nH_{2n} + n\; H_2O$$

$$n\,CO+(2n+1)H_2 \rightarrow C_nH_{2n+2}+nH_2O$$

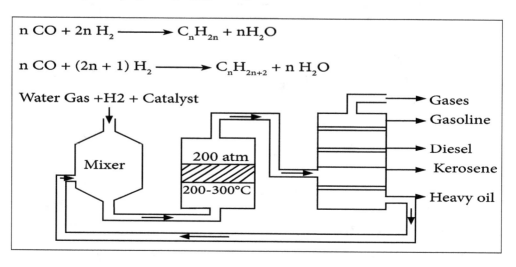

Reforming of Petrol

Reforming is a thermo catalytic process carried out to improve the octane number of petrol by bringing about changes in the structure of hydrocarbons. It involves a molecular rearrangement of hydrocarbons without any change in the number of carbon atoms to form new compounds It is usually brought about by passing the petroleum fraction at about 500°C over platinum coated on aluminum catalyst in the presence of hydrogen. The changes in structure could be isomerization, cyclization or aromatization.

Isomerization

Straight chain hydrocarbons with low octane number are converted to branched hydrocarbons having high octane number.

$$CH_3 - CH_2 - CH_2 - CH_2 - CH_2 - CH_2 - CH_3 \rightarrow CH_3 - CH - CH_2 - CH_2 - CH_2 - CH_3$$

n-heptane CH_3 Methyl hexane

Cyclization

Straight chain hydrocarbons with low octane number are converted to cyclic compounds having high octane number.

$$-CH_2-CH_2-CH_2- CH_2-CH_2-CH_3 \longrightarrow$$ methl cyclohexane $-CH_3 + H_2$

n- heptane   methl cyclohexane

Aromatization

Cyclic compounds with low octane number undergo dehydrogenation to form aromatic compounds having high octane number.

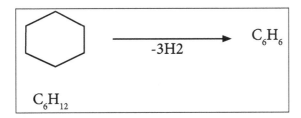

Methylcyclohexane   Toluene

Dehydrogenation: In dehydrogenation cyclo alkanes gets converted to aromatic hydrocarbon.

Knocking

The explosive combustion of petrol and air mixture produces shock waves in I.C. engine, which hit the walls of the cylinder and piston producing a rattling sound is known as knocking.

Mechanism of Knocking

Beyond a particular compression ratio the petrol mixture suddenly burns into flame. The rate of flame propagation increases from 20 to 25m/s to 2500m/s, which propagates very fast, producing a rattling sound.

The activated peroxide molecules decomposes to give number of gaseous products which produces thermal shock waves which hit the walls of the cylinder and piston causing a rattling sound which is known as knocking.

The reactions of normal and explosive combustion of fuel can be given as follows taking ethane as an example:

$$C_2H_6 + 3\frac{1}{2}\,O_2 \rightarrow 2CO_2 + 3H_2O \quad \text{(Normal comustion reaction)}$$
$$C_2H_6 + O_2 \rightarrow CH_3 - O - O - CH_3 \quad \text{(Explosive combustion reaction)}$$
$$CH_3 - O - O - CH_3 \rightarrow CH_3 - CHO + H_2O$$
$$CH_3 - CHO + 1\frac{1}{2}\,O_2 \rightarrow HCHO + CO_2 + H_2O$$
$$HCHO + O_2 \rightarrow CO_2 + H_2O$$

Petrol Knocking

Knocking is a something we want to try to prevent in petrol engines. It occurs in four stroke engines when some hydrocarbons ignite too early. It causes a knocking sound and reduces the engines performance.

We measure the problem using octane number. The higher the number, the better the fuel. A fuel with a low octane number is likely to cause knocking. One with an octane number over 100 should be fine.

There are two main ways of increasing the octane number of the petrol (i.e. improving it):

- Use alkanes with branching chains rather than straight chains.

- Use aromatic alkanes (with rings).

These alkanes are produced on the site using the isomeriser and the reformer. They are then blended into the petrol at end. The blended petrol is constantly monitored for its ocatane number.

Diesel Knocking

Cetane number is used to measure knocking in diesel engine.

Diesel knock is the clanking, rattling sound emitted from a running diesel engine. This noise is caused by the compression of air in the cylinders and the ignition of the fuel as it is injected into cylinder. This is much the same as a gasoline engine suffering from pre-ignition or spark knocking.

The timing of fuel being injected into the diesel engine is critical to prevent parts breakage, which can result from severe knock. A diesel engine functions differently than its gasoline counterpart. In a gasoline engine, the fuel is mixed with air and then compressed before an electric spark ignites the mixture. In a diesel engine, only air is compressed.

The fuel is then injected into the cylinder filled with compressed air and the heat from the compressed air ignites the fuel without the aid of an electric ignition. The telltale sound of an operating diesel engine is due in part to the fuel injection process.

By injecting raw fuel into extremely hot compressed air, the fuel ignites as the piston is still traveling up in the cylinder, causing a detonation and subsequent rattling sound to be heard. The process is compression driven and the higher the compression ratio within the cylinder, the greater the power output of engine.

While gasoline engines typically operate at 8:1 to 10:1 compression ratios on the street, the typical diesel engine operates at 14:1 to 25:1 compression ratios. This higher compression allows a diesel engine to operate much more efficiently than its gasoline. Diesel knock is a by-product of the raised compression and fuel injection process and is an acceptable result of the ignition sequence.

A diesel engine is difficult to start in cold weather due to its lack of an electronic ignition system. Many manufacturers equip diesel engines with glow plugs to aid in starting the engine in cold climates. A glow plug uses the battery to heat a wire coil red hot in the combustion chambers.

This causes more noticeable diesel knock in the engine until it reaches operating temperature. Knocking declines as the fuel begins to ignite more easily within the engine. Some manufacturers have created special engine mounts that help muffle diesel knock from passenger compartments.

As the cost of fuel rises, diesel engines are being fitted into an increasing amount of passenger vehicles due to superior fuel efficiency. Knock is seen by many as a tolerable side-effect of better fuel economy.

Octane Number of a Fuel

Octane number of a fuel is a measure of its ability to resist knocking. The knocking characteristics of petrol are described by the octane number. Higher the octane number lower is the knocking tendency & better is the quality of petrol.

The octane number is an arbitrary factor. Isooctane when used as a fuel found to have zero knocking hence its octane number is taken as 100. But when n-heptane was used as a fuel it had maximum knocking hence its octane number is taken as zero.

Octane number is equal to the percentage by volume of isooctane (2, 2, 4-trimethyl pentane) in a mixture of n-heptane and isooctane having the same knocking tendency compared to sample of gasoline being tested, isooctane has the best anti knocking properties and assigned an octane number of 100 whereas n-heptane has poor anti knocking property and assigned an octane number of zero.

The hydrocarbons present influence the knocking properties of gasoline which vary according to the series: straight chain paraffin > branched chain paraffin > olefin > cycloparaffin > aromatics. The fuel which has same knocking tendency with the mixture having 80% isooctane has octane number 80.

The most effective antiknock agent added is tetraethyl lead (TEL) along with ethylene dibromide which prevents the deposition of lead by forming volatile lead halides. Other anti knocking agents are tetra methyl lead (TML), tertiary butyl acetate, diethyl telluride.

1.0-1.5 ml of TEL is added per litre of petrol. TEL functions by being converted to a cloud of finely divided lead oxide particles, which react with any hydrocarbon peroxide molecules formed in the engine cylinder thereby solving down the chain oxidation re-action and preventing knocking.

Octane number is defined as "The percentage of isooctane present in a mixture of isooctane and n-hectane".

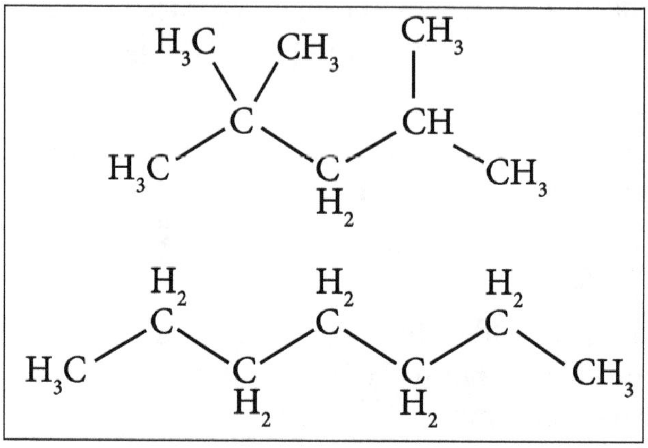

Octane number.

For a petrol sample under test whose octane number is to be determined, it is compared with reference standards of isooctane and n-heptane prepared at different ratios(90:10, 80:20, 75:25 etc) and the knocking characteristics of these is determined under same conditions as that of the sample under test.

Suppose the knocking characteristics of the fuel is same as that of 80:20 mixture, the octane number of the fuel is 80. Therefore the octane number is defined as the percentage by volume of isooctane present in a standard mixture of isooctane and n-heptane which has the same knocking characteristics as the petrol under test.

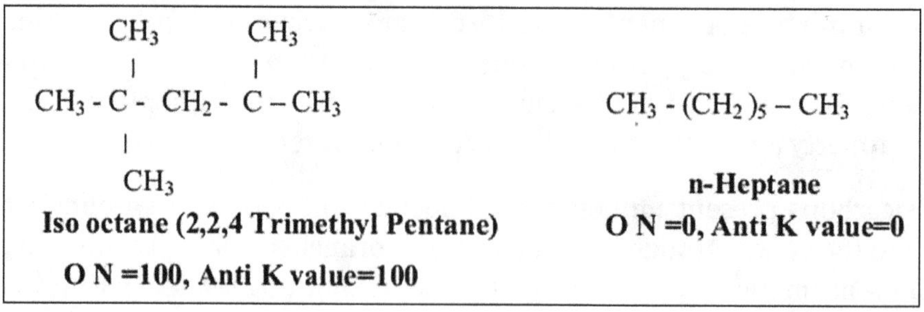

Iso octane (2,2,4 Trimethyl Pentane)

O N =100, Anti K value=100

CH_3 - $(CH_2)_5$ – CH_3

n-Heptane

O N =0, Anti K value=0

Cetane Number of a Fuel

Cetane number of a fuel is a measure of its ability to resist knocking. The knocking characteristics of diesel are described by the Cetane number, Cetane number is defined as the percentage by volume of Cetane(or hexa-decane) present in a standard mixture of cetane and α-methyl naphthalene which has the same knocking characteristics as the diesel fuel under test.

Generally diesel fuels with cetane numbers of 70-80 are used.

$C_{16}H_{34}$	$C_{10}H_7 - CH_3$
Cetane	α – Methyl napthalene
C N = 100, Anti K value = 100	C N = 0, Anti K value = 0

Prevention of Knocking (Anti-knocking Agents)

The substances added to gasoline to control knocking are called anti-knocking agents. Usually organometallic compounds are added to gasoline.

The followings are the substances that prevent knocking in I.C. Engines:

- TEL - Tetra Ethyl led(leaded petrol).
- TML - Tetra Methyl led.
- MTBE - Methyl Tertiary Butyl Ether(unleaded petrol)

Leaded Petrol

The petrol containing TEL or TML as the anti knocking agents is known as the leaded petrol. TEL or TML are the very good anti knocking agents but has some disadvantages as follows:

- When the combustion lead is deposited as lead oxide on piston and engine walls it leads to mechanical damage.
- The lead is a poisonous air pollutant.
- It spoils the catalyst used in catalytic converter.

Unleaded Petrol

The petrol, which contains anti-knocking agent other than lead, is known as unleaded petrol.

Example: MTBE is used, as an anti-knocking agent in place of TEL or TML and the petrol is known as unleaded petrol.

2.4   Power Alcohol

Alcohol is an excellent alternative motor fuel for gasoline engines. Alcohols are fuels belonging to the 'oxygenates' family having one or more oxygen which contributes to the combustion. Only two of the alcohols are technically and economically suitable as fuels for internal combustion engines, these are methanol and ethanol.

If about 20-25% ethyl alcohol is blended with petrol and is used as a fuel, it is known as power alcohol for internal combustion engine. Gasohol is a mixture of 90% unleaded gasoline and 10% ethyl alcohol (ethanol). Its performance as a motor vehicle fuel is comparable to that of 100% unleaded gasoline, with the added benefit of superior anti-knock properties (no premature fuel ignition).

No engine modifications are needed for the use of gasohol, which has in recent years gained some acceptance as an alternative to pure gasoline. Methanol is produced by a variety of process, the most common are: Distillation of wood, Distillation of coal, Natural gas and petroleum gas. Ethanol is produced mainly from biomass transformation or bioconversion.

Advantages of Power Alcohol

Alcohol has characteristics which makes it a natural engine fuel. For instance:

- It has a high "octane" rating, which prevents engine detonation (knock) under load.

- It burns clean, in fact, not only are noxious emissions drastically reduced, but the internal parts of the engine are purged of carbon and gum deposits which, of course, do not build up as long as alcohol is used as a fuel.

- An alcohol burning engine tends to run cooler than its gasoline-powered counterpart, thus extending engine life and reducing the chance of overheating.

- It has less starting problems.

- Alcohol removes all traces of moisture in the petrol.

2.4.1 Bio-diesel

Bio-diesel is a renewable, oxygenated fuel made from a variety of agricultural resources such as soybeans or rapeseeds. Bio-diesel is non-toxic, biodegradable replacement for petroleum diesel.

Bio-diesel is made from natural, renewable sources such as new and used vegetable (rape seed) oil and animal fats (tallow). Chemically bio diesel is described as a fatty acid mono alkyl ester.

Through a process called esterification, oils (rape seed) and fats are reacted with CH3OH and NaOH catalyst at 500°C to produce fatty acid methyl ester along with the co-products: glycerin, glycerin bottoms, soluble potash and soaps.

$$\text{Rape seed oil} + CH_3OH \rightarrow \text{bio-diesel} + \text{co-products}$$

Using bio-diesel in a conventional diesel engine substantially reduces emissions of unburnt hydrocarbons, carbon monoxide, sulphates, polycyclic hydrocarbons, nitrated polycyclic aromatic hydrocarbons.

One of the primary advantages of bio-diesel is its renewability. As a renewable, domestic energy source, bio-diesel can help reduce dependence on petroleum imports.

Bio-diesel is nontoxic, biodegradable and suitable for sensitive environments. Bio-diesel contains no petroleum but can be blended at any level with petroleum diesel to create a bio-diesel blend. Bio-diesel refers to the pure alternative fuel before blending with petroleum-based diesel fuel.

Advantage

- Substitute for petroleum which is non-renewable resources.

- Leads to reduction in C_2 emission by 75%.

- Complete burning of bio-diesel leading to improved emission.

- Biodegradable.

2.4.2 Gaseous fuels, Natural gas, LPG and CNG

Gaseous Fuels

Gaseous fuels occur in nature, besides being manufactured from solid and liquid fuels. Gas fuels are the most convenient because they require the least amount of handling and are used in the simplest and most maintenance-free burner systems.

Types of Gaseous Fuel

The following is a list of the types of gaseous fuel:

1. Fuels Naturally Found in Nature

- Natural gas.

- Methane from coal mines.

2. Fuel Gases Made from Solid Fuel

- Gases derived from coal.

- Gases derived from waste and biomass.

- From other industrial processes (blast furnace gas).

3. Gases made from Petroleum

- Liquefied Petroleum gas (LPG).

- Refinery gases.

- Gases from oil gasification.

4. Gases from some Fermentation Process

Advantages

Gaseous fuels due to the flexibility of their applications, shall posses the following advantages over solid or liquid fuels:

- It can be conveyed easily through pipelines to the actual place of need, thereby eliminating manual labour in transportation.

- It can be lighted at ease.

- It has high heat contents and hence help us in having higher temperatures.

- It can be pre-heated by the heat of hot waste gases, thereby affecting economy in heat.

- It's combustion can readily controlled for change in demand like oxidizing or reducing atmosphere, length flame, temperature, etc.

- It is clean in use.

- It does not require any special burner.

- It burns without any shoot or smoke and ashes.

- It is free from impurities found in solid and liquid fuels.

Disadvantages

- Very large storage tanks are needed.

- They are highly inflammable, so chances of fire hazards in their use is high.

Natural Gas

The natural gas is said to be a fossil fuel formed during layers of buried plants, gases and animals are exposed to intense heat and pressure over thousands of years. Therefore the energy that the plants originally obtain from the sun is stored in the form of chemical bonds in natural gas.

Characteristics of Natural Gas

Like crude oil, the natural gas is an energy source based on hydrocarbon chains, but the composition of natural gas is generally different than the composition of crude oil. The natural gas is primarily composed of methane, even though some natural gas deposits additionally contains substantial fractions of other hydrocarbon gases or liquids like ethane and propane. Most of the gas deposits also contain impurities like sulfur or other carbon compounds that should be separated prior to the gas being injected into transmission or distribution pipelines.

The gas deposits that consists primarily of methane are called as dry gas deposits, while those with larger fractions of other hydrocarbons are known as wet or rich gas deposits. Unlike oil, the natural gas is essential to its transportation system without pipelines, there is no economical way to get large quantities of gas to market.

Though, the natural gas pipelines generally require to be dedicated assets. Using oil or petroleum product pipelines to move the natural gas is not really possible and moving other products in natural gas pipelines is not possible without completely re-purposing the pipeline. This is an asset specificity and complementarity between natural gas and the pipeline transportation infrastructure has been a significant factor in the development of the natural gas market.

Chemical Composition of Natural Gas

The natural gas is primarily composed of methane, that also contains ethane, propane and also heavier hydrocarbons. It contains small amounts of nitrogen, carbon dioxide, hydrogen sulphide and also trace amounts of water.

Application

The application of gas cooling shall be seen in both Thailand and overseas for instance, the new Tokyo national airport in Japan, Kuala Lumpur International Airport and Suvarnabhumi Airport, Thailand.

The natural gas can be used for cooking by replacing the LPG in hotels, hospitals, restaurants and residences. The gas will be used with all types of stoves, ovens, grills and rice cookers.

The natural gas is used for producing the hot water and steam in hotels, the laundry services, the sterile process at hospitals and household residence.

LPG and CNG

CNG

It is stored in a high-pressure container usually at 3000 to 3600 psi. CNG is used mainly as an alternative fuel for internal combustion engines such as automobile engines. This generates low hydrocarbon emissions but a significant quantity of nitrogen oxide emissions.

LPG

The liquefied petroleum gas or liquid petroleum gas (LPG or LP gas) also referred to as simply propane or butane are the flammable mixtures of hydrocarbon gases which is used as fuel in heating appliances, cooking equipment and vehicles.

It is used as an aerosol propellant and also has a refrigerant replacing the chlorofluorocarbons in an effort to reduce damage to the ozone layer. It is specifically used as a vehicle fuel and often referred to as autogas.

2.5   Combustion

Combustion is a method where the rapid chemical combination of oxygen with the combustible elements of a fuel, resulting in the production of heat. Combustion is accomplished by mixing the fuel and air at elevated temperatures. The air supplies oxygen which unites chemically with the carbon, hydrogen and a few minor elements in the fuel to produce heat. Steam has been generated from the burning of a variety of fuels.

The heating value of a fuel may be determined either by a calculation from a chemical analysis or by burning a sample in a calorimeter.

The calculation should be based on an ultimate analysis that reduces the fuel to its elementary constituents of carbon, hydrogen, oxygen, nitrogen, sulphur, ash and moisture to secure a reasonable degree of accuracy.

The proximate analysis which determines only the percentage of moisture, volatile matter, fixed carbon and ash without determining the ultimate composition of the volatile matter cannot be used for computing the heat of combustion with the same degree of accuracy as an ultimate analysis but estimates may be based on the ultimate analysis which is fairly correct.

Combustion refers to rapid oxidation of the fuel accompanied by the production of heat or heat and light. Complete combustion of a fuel is possible in the presence of an adequate supply of oxygen only.

Oxygen (O_2) is one of the most common elements on the earth making up to 20.9% of our air. Rapid fuel oxidation results in large amounts of heat. Solid or liquid fuels must be changed to gas before they burn. Generally heat is required to change liquids or solids into gases. Fuel gases will burn in their normal state if enough air is present.

Most of the 79% of air is nitrogen, with traces of other elements. Nitrogen is considered to be a temperature reducing dilutant that must be present to obtain the oxygen required for combustion.

Nitrogen reduces the combustion efficiency by absorbing heat from the combustion of fuels and diluting the flue gases. This reduces heat available for transfer through the heat exchange surfaces. It also increases the volume of combustion by-products, which have to travel through the heat exchanger and up the stack faster to allow the introduction of the additional fuel air mixture.

This nitrogen can combine with oxygen (at high flame temperatures) to produce oxides of nitrogen (NO_x), which are known as toxic pollutants.

Carbon, sulphur and hydrogen in the fuel combine with oxygen in the air to form carbon-dioxide, sulphur dioxide and water vapour, releasing 8084 kcals, 2224 kcals & 2224 kcals of heat respectively.

Under certain conditions, carbon may also combine with Oxygen to form Carbon Monoxide, which results in the release of a smaller quantity of heat (2430 kcals/kg of carbon) Carbon burned to CO_2 will produce more heat per pound of fuel than when CO or smoke are produced.

$$C + O_2 \rightarrow CO_2 + 8084 \text{ kCals/kg of Carbon}$$

$$2C + O_2 \rightarrow 2CO + 2430 \text{ kCal/kg of Carbon}$$

$$2H_2 + O2 \rightarrow 2H_2O + 28,922 \text{ kCal/kg of Hydrogen}$$

$$S + O_2 \rightarrow SO_2 + 2,224 \text{ kCal/kg of Sulphur}$$

Each kilogram of CO formed means a loss of 5654 kCal of heat.

2.5.1 Calculation of Air for the Combustion of a Fuel

Problems

1. Let us calculate the mass of air needed for complete combustion of 5kg of coal containing H= 15%. C = 80%, O = rest.

Solution:

5 kg of coal contains: C = 4 kg, H = 0.75 kg, o = (5 − 4 − 0.75) kg = 0.25 kg.

Amount of air required for complete combustion of 5 kg coal

= [5 x (32/12) + 0.75 x (16/2) − 0.25] kg x (100/23)

= [13.333 + 6.000-0.25] kg x (100/23) = 82.97 kg

2. A sample of coal was found to contain C − 80%, H − 5%, o − 1%, N − 2% remaining

being ash. Let us calculate the amount of minimum air required for complete combustion of 1 kg of coal sample.

Solution:

Combustion reaction	Weight of air required
$C + O_2 \rightarrow CO_2$ 12 32	800g (32/12) = 2146 gm 50g(16/2) = 400 gm
$2H + 0.5O_2 \rightarrow H_2O$ 2 16	Total = 2546 gm Less O in fuel = 10 gm Net O_2 required =2536 gm

Weight of air required = 2536 g(100/23) = 11026 gm = 11.026kg.

3. Let us calculate the weight and volume of air required for the combustion of one kg of Carbon.

Solution:

Carbon undergoes combustion according to the equation.

$$C + O_2 \rightarrow CO_2$$

12 32

Thus weight of O_2 required for combustion of 12 gm of C = 32 gm.

Hence weight of oxygen required by 1 kg of carbon $= \dfrac{32}{12} \times 1 = 2.667$ kg

Weight of air(containing 23% oxygen)required $= \dfrac{100}{23} \times 2.667 = 11.59$ kg

Now since 32 gm of oxygen occupies 22.4 liters at NTP

$$\therefore 2.667 \times 1000 \text{ gm of } O_2 \text{ will occupy} = \dfrac{22.4}{32} \times 2.667 \times 1000 = 1866.9 \text{ L}$$

So, volume of air (containing 21% oxygen) required

$$= \dfrac{100}{21} \times 1866.9 = 8890 \text{ Litres} = 8.89 \text{ m}^3$$

2.5.2 Flue Gas Analysis, Orsat Apparatus

Flue Gas Analysis

The mixture of gases like CO_2, CO, O_2 etc., coming out from the combustion chamber is called as flue gas. The amount of flue gases like CO_2, CO, O_2 etc., can be estimated by using Orsat's apparatus on the absorption principle.

The gases like CO_2, CO and O_2 are absorbed by KOH, alkaline pyro-gallon and ammo-niacal cuprous chloride solutions respectively.

- High CO in the flue gas shows incomplete combustion of the fuel and short supply of oxygen.

- High CO_2 and O_2 in the flue gas shows complete combustions of the fuel and excess supply of oxygen.

It consists of a horizontal tube having three way stop cock at one end. The end of three way stop cock is connected to a U tube containing fused $CaCl_2$ to remove moisture in the gas. The end of the tube is connected with a graduated burette. The burette is sur-rounded by a water jacket in order to keep the temperature of gas constant.

The lower end of the burette is connected by a water reservoir by means of rubber tube. The level of the water in burette can be raised or lowered by raising or lowering the reservoir. The middle of the horizontal tube is connected with 3 bulbs (A, B and C) for absorbing flue gases as follows:

- Bulb 'A' containing KOH solution and it absorbs only CO_2.

- Bulb 'B' containing alkaline pyro-gallon solution and it absorbs only O_2.

- Bulb 'C' containing ammoniacal cuprous chloride solution and it absorbs CO.

Working of Orsat Apparatus

Now the three way stop cock is opened and the burette is filled with water by raising the water reservoir to remove air from the burette. Then the flue gas is taken in the burette up to 100 cc by raising and lowering the reservoir. The 3 way stop cock is now closed.

Absorption of Gases on Bulbs

1. Absorption of CO_2

The stopper of the bulb 'A' is opened and the flue gas is allowed to pass by raising the water reservoir. CO_2 present in flue gas is absorbed by KOH. This process is repeated several times by raising and lowering the water reservoir until the volume of burette becomes constant. The decrease in volume of burette indicates the volume of CO_2 in 100 cc of the flue gas. Now the stopper of the bulb 'A' is closed.

2. Absorption of O_2

The stopper of the bulb 'B' is opened and the flue gas is allowed to pass. O_2 present in the flue gas is absorbed by alkaline pyrogallol. This process is repeated several times until the volume of burette becomes constant. The decrease in volume of flue gas in burette indicates the volume of O_2. Now the stopper is closed.

3. Absorption of CO

The stopper of the bulb 'C' is opened and the flue gas is allowed to pass. 'CO' present in the flue gas is absorbed by ammoniacal cuprous chloride solution. This process is repeated several times until the volume of burette becomes constant. The decrease in volume of flue gas in burette indicates the volume of CO. The remaining gas in the burette after the absorption of CO_2, O_2 and CO is taken as nitrogen.

The % of $N_2 = [100 - (\%\ of\ CO_2 + \%\ of\ O_2 + \%\ of\ CO)]$

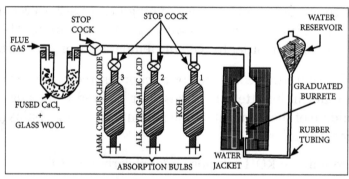

Flue Gas Analysis.

2.5.3 Numerical Problems on Combustion

1. Let us calculate the volume of air required for complete combustion of 1 m³ of gaseous fuel having the composition: CO = 46%, CH_4 = 10%, H_2 = 4%, C_2H_4 = 2%, N_2 = 1% and the remaining being CO_2.

Solution:

Given:

   1 m³ of fuel contains:

   4/100 = 0.04 m³ of H_2.

   10/100 = 0.10 m³ of CH_4

   1/100 = 0.01 m³ of N_2

   46/100 = 0.46 m³ of CO

   2/100 = 0.02 m³ of C_2H_4

N_2 and CO_2 are non-combustible constituents they do not undergo combustion.

The combustion equation is written as follows:

(a) $H_2 + 1/2\ O_2 \rightarrow H_2O$

1 vol 0.5 vol.

1   of H_2 requires 0.5 m^3 of O_2

0. 04 m^3 of H_2 requires = $0.5 \times 0.04/1 = 0.02$ m^3 of O_2

(b) $CH_4 + 2O_2 \rightarrow CO_2 + 2H_2O$

1 vol 2 vol.

1 m^3 of CH_4 requires 2 m^3 of O_2

0. 10 m^3 of CH_4 requires = $0.10 \times 2/1 = 0.2$ m^3 of O_2

(c) $CO + 1/2\ O_2 \rightarrow CO_2$

1 vol 0.5 vol.

1 m^3 of CO requires 0.5 m^3 of O_2

0.46 m^3 of CO requires = $0.46 \times 0.5/1 = 0.23$ m^3 of O_2

(d) $C_2H_4 + 3O_2 \rightarrow 2CO_2 + 2H_2O$

1 vol 3 vol.

1 m^3 of C_2H_4 requires 3 m^3 of O_2

0.02 m^3 of C_2H_4 requires = $0.02 \times 3/1 = 0.06$ m^3 of O_2

Total volume of O_2 required = 0.02 + 0.2 + 0.23 + 0.06

= 0.51 m^3 of O_2

We know that, 21 m^3 of O_2 is supplied by 100 m^3 of air

0.51 m^3 of O_2 is supplied by = $100 \times 0.51/21 = 2.428$ m^3 of air

Amount of oxygen required my 100 m^3 of fuel = 2.428 m^3 of air

2. A fuel contains C = 75%, H = 4%, O = 5%, S = 7% remaining ash. Let us calculate the minimum quantity of air required for complete combustion of 1kg of fuel.

Solution:

Given:

Weight of the fuel = 1 kg

Weight of C in the fuel = 0.75 kg

Weight of H in the fuel = 0.04 kg

Weight of O in the fuel = 0.05 kg

Weight of S in the fuel = 0.07 kg

(i) $CO + O_2 \rightarrow CO_2$

12 kg of carbon requires 32 kg of oxygen

0.75 kg carbon requires = (32/12) x 0.75 = 2 kg

(ii) $H_2 + 1/2\, O_2 \rightarrow H_2O$

2 kg of hydrogen requires 16 kg of oxygen

0.04 kg hydrogen requires = (16/2) x 0.04 = 0.32 kg

(iii) $S + O_2 \rightarrow SO_2$

32 kg of carbon requires 32 kg of oxygen

0.07 kg carbon requires = (32/32) x 0.07 = 0.07 kg

Total weight of oxygen required = 2 + 0.32 + 0.07 = 2.39 kg

But weight of oxygen already present = 0.05 kg

Actual weight of oxygen required = 2.39-0.05 = 2.34 kg

Weight of air required = 2.34 x (100/23) = 10.17 kg of air

2.6   Explosives: Rocket Fuels

When a substance or a mixture is subjected to thermal or mechanical shock, it gets very rapidly oxidized exothermically into products of greatly increased volume, with a sudden release of potential energy called explosive.

Power to weight ratio of an explosive is nothing but the amount of power available from a given weight (or volume) of explosive.

Characteristics of Explosives

- It should be cheap and stable under normal conditions.

- It must have at least one chemical bond that can be easily broken or the molecule should have a low energy of dissociation.

- The rate of decomposition should be fast to produce a large volume of gaseous products exothermically.

- It should have a positive oxygen-balance. The oxygen-balance is expressed as a percentage surplus or deficiency of oxygen by weight. The oxygen balance indicates the oxygen contained in the molecule, which can be utilized to oxidize the C and H to CO_2 and H_2O respectively.

Classification of Explosives

Explosives are usually classified into three broad groups as given below:

- Primary explosives (detonators).
- Low explosives (propellants).
- High explosives.

Uses of Explosives

Explosives can be used for constructive as well as destructive purposes. They can be used:

- For blasting of ores of iron and other metals.
- For breaking down coal, quarrying granite, limestone etc.
- In formation of tunnels, roads etc. which require blasting.
- As an ammunition in was.
- To manufacture bombs, grenades etc.
- For launching satellites with the help of rockets.

Blasting Fuse

A fuse is a thin waterproof canvas length of tube containing gunpowder arranged to burn at a given speed for setting off charges of explosives.

They are of two types, safety fuse and detonating fuse:

- Safety fuse: It is employed in initiating caps where electrical firing is not used.
- Detonating: It has a velocity of over 6,000 m per second and consists of a charge of high velocity explosive, such as TNT, contained in a small diameter bent tube.

Manufacture of some Explosives

Trinitrotoluene (TNT)

Nitration of toluene by nitrating agent (mixture of conc. HNO_3 and conc. H_2SO_4 in 1:1 ratio) in a reaction tank.

The product is washed with ammonical solution of Na2SO3 and then with cold water, when TNT crystallizes out. Crystals of TNT are filtered and purified by melting.

Nitroglycerine

Prepared by reaction of glycerol with conc. H2SO4 (60%) and conc.HNO3 (40%) at 10°C with constant stirring.

Applications of Explosives

Objects at a smaller distance can be shattered with high brisance explosives: explosives with low brisance can be used as propellants. i.e.. shells, bombs and mines need high brissance explosives while all propellants belong to low-explosives category.

For sub-soil blasting, the explosives are burst a few feet below the land. In this process, the sub-soil is brought up to the surface and the liberated gases SO_2, CO, etc., can kill insects. If the explosives are nitrate-based, some fertilizers also get incorporated into the soil.

An explosive attached to a piece of metal is used to shape the metal to required shapes that are not achievable by other methods. This kind of metal-forming by explosives is of recent origin.

The explosives for dislodging the coal seam in coal mines must be carefully chosen. A long-lasting flame may cause dangerous explosions due to the exothermic burning of methane and coal dust. The addition of NaCI and AN to GTN not only helps to lower the flame temperature but also the duration of the flame. The mixture, known as permitted explosive, is used in coal mines.

Explosive Range

Explosive range is the concentration of a substance in air between the LEL (lower explosive limit) and UEL (upper explosive limit). In this range the substance will readily ignite.

Lower Explosive Limit

Lower Explosive Limit (LEL) is the minimum concentration (% in air) of a substance in air which is required for ignition. Concentrations below the LEL will not ignite. Below the LEL, the mixture is called "lean". If the concentration of a flammable vapor or gas is greater than 10% of the LEL evacuate the area.

Upper Explosive Limit

Upper Explosive Limit (UEL) is the maximum concentration (% in air) of a substance in air which is required for ignition. Concentration above the UEL will not ignite. Above the UEL, the mixture is called "rich".

Rocket Fuels

Modern rockets come in two main categories, solid fuel and liquid fuel.

Liquid- rocket Fuel

Liquid-fuel rockets most commonly use liquid oxygen and either kerosene or liquid hydrogen. These combinations work well in space and down closer to the ground, resulting in a multitude of uses from the first stages of the Saturn V and Falcon rockets to the Space Shuttle's main engines used to get the orbiter in position in space.

Advantages

- Variable thrust- the amount of fuel and rate of burn can be changed in flight.
- Liquid-fuel boosters are more easily re-usable.

Disadvantages

- Fragile, many complex parts.
- Oxidizer (liquid oxygen) must be kept extremely cold.

Solid-rocket Fuel

Solid-rocket fuel is easier and cheaper to handle and make as we don't have to cool materials to cryogenic temperatures. This means solid fuel is quite prevalent in military applications and also the first stages of space rockets. Made from powdered aluminium and an oxidizer, this fuel is often used in booster rockets to give craft that extra kick needed to lift off and make their way into space.

Advantages

- Very stable, durable.

- More thrust for a similar size rocket.

Disadvantages

- Can't be turned off- once the burn starts, it goes until fuel is used up.

- Fuel decomposes, must be replaced.

3

Electrochemical Cells and Corrosion

3.1 Galvanic Cells, Reversible and Irreversible Cells, and Single Electrode Potential

Galvanic Cells

Galvanic cells are electrochemical cells in which the electrons are transferred due to redox reaction is converted to electrical energy. The flashlight battery is an electrochemical cell. Electrochemical cell is something that changes chemical energy to electrical energy. It contains two compartments each with an electrode submerged in an electrolyte.

The electrode is just a conductor usually a metal that connects to a non-metallic part of a circuit. The electrolyte is a fluid that conducts electricity. One of the plates is positive and one of them is negative. Those two plates are called the cathode and the anode and they are connected by a wire that completes the circuit.

Often one metal strip in the battery is copper and it acts as a cathode which is the electrode where, reduction takes place and electrons are gained. The other electrode is the anode often zinc where, oxidation takes place and electrons are lost. Electrons leave the system from the anode and go into the wire.

The electrons move from the anode to the cathode through the wire. So what happens is the electrons on the anode are lost and travels through the wire to the cathode where electrons are gained. The solution of the two electrodes helps in the conduction of the electrons from the surface of the electrode to the wire.

The two electrodes are in separate compartments separated by a porous barrier or salt bridge. The barrier or bridge allows the ions in both solutions to move from one side to the other to prevent charge from building up on the electrodes.

The chemical reaction that occurs in electrochemical cell is a redox or reduction-oxidation reaction. As its name implies that reaction's made up of two parts: a reduction reaction where electrons are gained and an oxidation reaction where electrons are lost.

They are electrochemical cells in which the electrons transferred due to redox reaction, are converted into electrical energy. A galvanic cell consists of two half-cells with each half-cell contains an electrode.

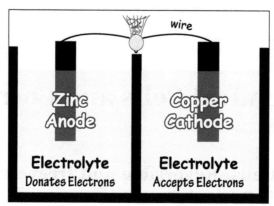

Electrochemical cell.

The electrode at which oxidation takes place is called anode and the electrode at which reduction occurs is called cathode. The electrons liberated to the electrolyte from the metal leaves the metal ions at anode.

The electrons from the solution are accepted by cathode metal ion to become metal. Galvanic cell is generally represented as follows:

$$M_1/M_1^+ \, || \, M_2^+/M_2 \text{ or } M_1/\left(\text{Salt of } M_1\right) || \, M_2/\left(\text{Salt of } M_2\right)$$

Where, M_1 & M_2 are anode and cathode respectively and M_1^+ & M_2^+ are the metal ions in the respective electrolyte. The symbol $||$ denotes salt bridge.

The above representation of galvanic cell is known as galvanic cell diagram.

Example: The typical example for galvanic cell is Daniel cell.

Setting Up Galvanic Cells

Galvanic cells harness the electrical energy available from the electron transfer during a redox reaction to perform useful electrical work. The key to gathering the electron flow is to separate the oxidation and reduction half-reactions, connecting them by a wire, so that the electrons should flow through that wire. That electron flow, known as a current, may be sent through a circuit which could be part of any number of electrical devices like radios, televisions, watches, etc.

The figure below shows two typical setups for galvanic cells. The left hand cell diagram shows an oxidation and a reduction half cell reaction joined by both a wire and a porous disk, while the right hand cell diagram shows the same cell substituting a salt bridge for the porous disk.

The salt bridge or porous disk is important to maintain the charge neutrality of each half-cell by permitting the flow of ions with minimal mixing of the half-cell solutions. As electrons are transferred from oxidation half-cell to the reduction half-cell, a negative charge builds in the reduction half-cell and a positive charge in the oxidation half-cell.

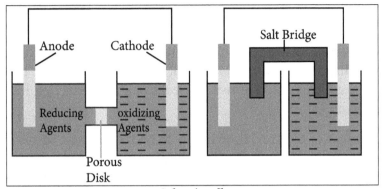

Galvanic cell.

The charge buildup would serve to oppose the current from anode to cathode effectively stopping the electron flow if the cell lacked a path for ions to flow between the two solutions. The above figure points out that the electrode in the oxidation half-cell is known as the anode and the electrode in the reduction half-cell is known as the cathode.

The anode, as it is the source of the negatively charged electrons it is typically marked with a minus sign (−) and the cathode is marked with a plus sign (+).

Physicists define the direction of current flow as the flow of positive charge based on an 18th century understanding of electricity. As we currently know, negatively charged electrons flow in a wire. Therefore, chemists indicate only the direction of electron flow on cell diagrams and not the direction of current. To make that point clear, the direction of electron flow is indicated on with a arrow and symbol for an electron, e^-.

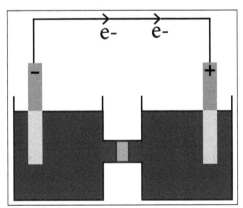

Galvanic cell showing direction of electron flow.

Electrolytic Cell

Voltaic cells use a spontaneous chemical reaction to drive an electric current through an external circuit. These cells are important because they are the basis for the batteries. But they aren't the only kind of electrochemical cell. It is also possible to construct a cell that does work on a chemical system by driving an electric current through the system.

These cells are called electrolytic cells. Electrolysis is used to drive an oxidation-reduction reaction in a direction in which it does not occur spontaneously. Electrolytic cells, like galvanic cells, are composed of two half-cells one is a reduction half-cell, other is an oxidation half-cell.

Though the direction of electron flow in electrolytic cells may be reversed from the direction of spontaneous electron flow in galvanic cells, the definition of both cathode and anode remain the same reduction takes place at the cathode and oxidation occurs at the anode.

Because the directions of both half-reactions have been reversed, the sign, but not the magnitude, of the cell potential has been reversed. Note that copper is spontaneously plated onto the copper cathode in the galvanic cell whereas it requires a voltage greater than 0.78 V from the battery to plate iron on its cathode in the electrolytic cell.

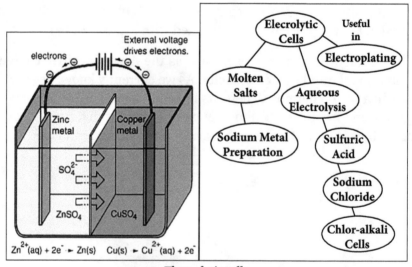

Electrolytic cell.

If a voltage greater than 1.10 volts is applied to a cell under standard conditions, then the reaction will be driven by removing Cu from the copper electrode and plating zinc on the zinc electrode.

$$Cu(s)+Zn_2+(aq) -> Zn(s)+Cu_2+(aq)$$

Electrolytic processes are very important for the preparation of pure substances like aluminum and chlorine.

Reversible and Irreversible Cells

A cell is said to be reversible if it fulfills the following conditions:

- Suppose the EMF of the cell is E_{cell}. If an external EMF of the same magnitude is applied to balance the EMF of the cell no current should flow to and from the cell.

- If the external EMF is decreased by an infinitesimal amount dE_{cell}, a small current should flow from the cell and a chemical reaction, proportional to the quantity of electricity passing, should take place in cell.

- If the external EMF is increased by an infinitesimal amount dE_{cell}, the current should flow in the opposite direction, proportional to the quantity of electricity passing and the cell reaction should be reversed.

If the cell does not satisfy these conditions, it is said to be irreversible.

Single Electrode Potential

It is not possible to measure the absolute potential of a single electrode because neither the oxidation nor the reduction reaction can occur by itself. Moreover, we need two electrodes to measure the potential difference between two points.

This difficulty can be solved selecting one of the electrodes as a reference electrode and arbitrarily fixing the potential of that electrode as zero. Thus, it is possible to obtain the potential of an electrode, if the given electrode is coupled with another electrode having a potential of zero volts. The electrode whose potential can be taken as zero volt is called as primary reference electrode.

A standard hydrogen electrode has been assigned an electrode potential of 0.00 volt at all temperatures and is used as a primary reference electrode. The electrode potential of any other electrode is obtained by coupling it with a standard hydrogen electrode.

The potential of the given electrode may be negative or positive with respect to standard hydrogen electrode potential. The electrode potential represents the electron-accepting power of an electrode relative to that of the standard hydrogen electrode, i.e., the electrode potentials are reduction potentials.

The electrode potential depends upon:

- The nature of the metal and its ions.

- Concentration of the ions in the solution.

- Temperature.

Since the potential depends upon the concentration of the ions in solution we adopt the standard as 1 mole of ions per liter of solution. Electrode potentials under these standard conditions are called standard electrode potentials and are denoted as $E°$. The standard conditions are:

- The concentration of the ions that take part in the electrode reaction should be 1 mol L^{-1}.

- For all the gaseous substances, which take part in electrode reaction, the pressure should be equal to 1 atm.

- A temperature of 298 K.

Nernst Equation

The Nernst Equation is used to determine single electrode potential of the cell. The Nernst Equation is derived from the emf and the Gibbs energy under non-standard conditions. Under standard conditions, the Gibbs free energy equation is then,

$$\Delta G° = -nFE°$$

Since,

$$\Delta G = AG° + RT \ln K \quad ...(1)$$

Substituting $\Delta G = -nFE$ and $\Delta G° = -nFE°$ into equation (1) we have,

$$-nFE = -nFE° + RT \ln K$$

Divide both sides of the equation by $-nF$ we have,

$$E = E° - RT/nF \ln K \quad ...(2)$$

Equation (2) can be rewritten in the form of log base 10, When ln is converted to log by 2.303 then,

$$E = E° - 2.303 RT/nF \log K \quad ...(3)$$

At standard temperature T=298K, the R is 8.314, F is 96,500 coulombs which is equal to 0.0592 V, so equation (3) turns into,

$$E = E° - 0.0592/n \log K.$$

The equation above indicates that the electrical potential of a cell depends upon the reaction quotient K of the reaction. As the redox reaction proceeds reactants are consumed thus concentration of reactants decreases.

Conversely the products concentration increases due to the increased products formation. As this happens cell potential gradually decreases until the reaction is at equilibrium at which $\Delta G = 0$.

Significance

1. Calculate the EMF of the cell.

$$\text{Zn}(s) \mid \text{Zn}^{2+}(0.024 \text{ M}) \mid\mid \text{Zn}^{2+}(2.4 \text{ M}) \mid \text{Zn}(s)$$

Understandably the Z^{2+} ions try to move from the concentrated half cell to a dilute solution. That driving force gives rise to 0. 0592 V.

At equilibrium concentrations of the two half cells will have to be equal in which case the voltage will be zero.

2. The equilibrium constant K may be calculated using standard cell potential $E°$ for the reaction.

3. From the standard cell potentials we can calculate the solubility product for the following reaction.

$$\text{AgCl} = \text{Ag}^+ + \text{Cl}^{-1}$$

Nernst Equation

Consider the following redox reaction,

$$\text{Mn} + \text{ne} \rightleftharpoons \text{M}$$

For such a redox reversible reaction the free energy change (ΔG) and its equilibrium constant (K) are interrelated as,

$$\Delta G = -\text{RTl n K} + \text{RTl n} \frac{[\text{Product}]}{[\text{Reactant}]}$$

$$= \Delta G° + \text{RTl n} \frac{[\text{Product}]}{[\text{Reactant}]} \qquad ...(1)$$

Where,

$\Delta G°$ = Standard free energy change. The equation (1) is known as Van't Hoff isotherm.

The decrease in free energy $(-\Delta G)$ in the above reaction will produce electrical energy. In the cell if the reaction involves transfer of 'n' number of electrons the 'n' Faraday of electricity will flow. If E is the emf of the cell, then the total electrical energy (nEF) produced in the cell is,

$$-\Delta G = nEF$$

Or,

$$-\Delta G^\circ = nEF \quad ...(2)$$

Where,

$-\Delta G$ = Decrease in free energy change.

$-\Delta G^\circ$ = Decrease in standard free energy change.

Comparing equation (1) and (2) it becomes,

$$-nEF = -nE^\circ F + RT\ \text{In} \frac{[M]}{[M^{n+}]} \quad ...(3)$$

Dividing the above equation (3) by -nF,

$$E = E^\circ - \frac{RT}{nF} \ln \frac{[M]}{[M^{n+}]}$$

[Activity of solid metal [M] = 1]

$$E = E^\circ - \frac{RT}{nF} \ln \frac{1}{[M^{n+}]}$$

$$= E^\circ + \frac{RT}{nF} \ln [M^{n+}]$$

$$E = E^\circ - \frac{RT}{nF} \ln \frac{[Product]}{[Reactant]}$$

$$E = E^\circ + \frac{RT}{nF}.2.303 \log [M^{n+}] \quad ...(4)$$

When, R = 8.314 JK-1 mole-1

F = 96,500 coulomb

T = 298 k

$$E = E^\circ_{red} + \frac{0.0591}{n} \log [M^{af}] \quad ...(5)$$

Similarly for oxidation potential,

$$E = E^o_{n\infty} + \frac{0.0591}{n} \log\left[M^{af}\right] \quad ...(6)$$

The above equation (5) and (6) are known as "Nernst equation for single electrode potential".

3.2   Electro Chemical Series and uses of this Series

Electrochemical Series

The electrode potentials of various electrodes are arranged in the order of increasing standard reduction potentials with respective to hydrogen scale. This arrangement is known as electrochemical series.

Importance of electrochemical series:

1. Calculation of Standard EMF of a Cell

The standard EMF of a cell can be calculated using the standard reduction potentials of right hand side and left hand side electrodes from the emf series,

$$E^o_{cell} = E^o_R - E^o_L$$

EMF

"EMF of a cell is defined as the algebraic difference between the reduction potentials of the cathode and the anode".

The measured EMF of a cell is,

$$E_{cell} = E_{cathode} - E_{anode}$$

$$E_{cell} = E_{right} - E_{left}$$

Where $E_{cathode}\left(E_{right}\right)$ and $E_{anode}\left(E_{left}\right)$ are the reduction electrode potentials of the cathode and anode respectively.

2. Relative Ease of Oxidation and Reduction

Higher the value of standard reduction potential greater is the tendency to get reduced. For example in the electrochemical series fluorine has high positive value of reduction potential (+2.87 V) and is reduced. Lower the value of standard reduction potential greater is the tendency to get oxidized. In the electrochemical series lithium has the lower negative reduction potential value (-3.05V) and is oxidized.

3. Predicting Spontaneity of a Reaction

The spontaneity of a reaction can be determined from the standard emf of the cell. If the emf of a cell is positive, the reaction is spontaneous. If it is negative the reaction is non-spontaneous.

$$E^o_{cell} = E^o_r - E^o_l = +ve \text{ (spontaneous).}$$

$$E^o_{cell} = E^o_r - E^o_l = -ve \text{ (non spontaneous).}$$

4. Displacement Behaviour of Hydrogen

If zinc electrode is immersed in H_2SO_4 solution it displaces hydrogen from acid solution.

$$Z_n + H_2SO_4 \rightarrow Z_nSO_4 + H_2^-$$

If silver electrode is immersed in H_2SO_4 solution it will not displace hydrogen from acid solution because it has higher reduction potential value.

$$Ag + H_2SO_4 \rightarrow \text{No reaction}$$

Use of EMF Series

- Calculation of standard emf of the cell: $E^o_{Cell} = E^o \text{ (R.H.E - E°L.H.E).}$

- Relative ease of oxidation or reduction: Higher the positive value of E^o greater is tendency to get reduced whereas higher is negative value of E^o greater is tendency to get oxidized.

- Displacement of one element by the other Metal which lie higher in the series can displace those elements which lie below them in the series.

- Determination of equilibrium constant for the reaction,

    $$-\Delta G^o = RT \ln K = 2.303 \, RT \log K$$

    $$\log K = -\nabla G^o / 2.303 \, RT$$

    since $-\nabla G^o = nFE^o$

$$\log K = nFE°/2.303RT$$

- Hydrogen displacement behavior.

Metal placed above H2 in the emf series will displace hydrogen from an acid solution.

$$Zn + H_2SO_4 \rightarrow ZnSO_4 + H_2$$

$$E°Zn = -0.76V$$

$$Ag + H_2SO_4 \rightarrow \text{ no reaction}$$

$$E°MAg = + 0.80V$$

Predicting spontaneity of a reaction: If E° is positive the cell reaction is spontaneous & if E° is negative the cell reaction is not feasible.

3.2.1 Standard Electrodes: Hydrogen Electrode -Calomel Electrode

Standard Electrodes

It is a tendency of a metal to loose or gain electron when it is immersed in its own solution at 25°C and 1M Concentration.

Example: $Zn/ZnSO_4$ (1M) at 25°C.

When we build a cell to measure a potential of a chemical reaction, each pole of the cell contains an electrode that transfers electrons from the solution to the wire or vice versa.

There are two main types of electrodes:

- Indicator electrodes - Electrodes that respond to an analyte concentration.
- Reference electrodes - Electrodes that maintain a fixed potential.

An indicator electrode may be as simple as a piece of Pt wire that is inert, so it doesn't take part in a reaction but simply provides a path for the electrons to flow. Many times an indicator electrode may be made of a metal that takes part in the ½ reaction so indicator electrodes can be simple or more complicated.

A best example is the Standard Hydrogen Electrode.

Standard Hydrogen Electrode

This is a great reference electrode because we know its potential is 0 so there is no built in potential that we have to compensate for. There are two widely used alternative electrode that are much safer to use.

Silver-Silver Chloride electrode,

$$AgCl(s) + e^- \rightarrow Ag(s) + Cl^-$$

The potential of this electrode is based on the ½ reaction and it has potentials of +.222 V if we use 1M KCL or .197 V is used for saturated KCL. Since both Ag and AgCl are solids the only thing that will vary voltage of this cell is Cl hence the two different potentials.

The saturated KCL is more reproducible in the lab because we just add KCl to the electrode until we see crystals and we don't need to get the concentrations exactly as 1M. The silver-Silver Chloride is slightly expensive to make, so there is a less expensive alternative.

The Calomel Electrode

The potential for this electrode is governed by the equation,

$$1/2\,Hg_2\,Cl_2(s) + e^- \rightarrow Hg(l) + Cl^-$$

Again all but Cl are solids so we get 2 potentials +.268 if KCl=1M +.241 if KCl is saturated. Given a potential measured against any one of these three common reference electrodes we should be able to convert the potential to measure against one of the other electrodes.

For example we measured a new potential gains of an SHE electrode and found it has a +.2V potential then the potential we measured would be against a Saturated Calomel Electrode.

Our measured potential is +.2 relative to 0 and the SHE is at +.241 from 0, so using this electrode we would measure a +.041 (.241-.2) potential. We measure a potential of +.1 relative to a SHE electrode then the measure relative to a Ag/AgCl electrode is,

.241 + .1 = .341

.341-.197 = .144

Measurement of Electrode Potential using Calomel Electrode (CE)

In order to measure the potential of a given electrode using calomel electrode, a galvanic cell is obtained by combining both the electrodes according to cell conventions. If a given electrode undergoes oxidation w.r.t. CE, then it is taken as anode and CE is taken as cathode.

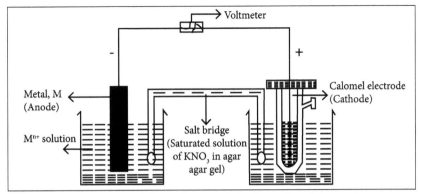

Measurement of electrode potential using calomel electrode.

Galvanic cell obtained in this case is represented as,

$$M \mid M+n \parallel KCl \mid Hg2Cl2(s) \mid Hg(l)$$

At anode, oxidation occurs, $M \rightarrow M+n+ne^-$

At cathode, reduction occurs, $Hg_2Cl_2 + 2e^- \rightarrow 2\,Hg + 2Cl^-$

EMF of cell, $E_{Cell} = E_{cathode} - E_{Anode}$

$$E_{Cell} = ECE - EMn+/M$$

$$E_{Mn}+/M = ECE - E_{cell}$$

If a given electrode undergoes reduction w.r.t. CE, then it is taken as cathode and CE is taken as anode.

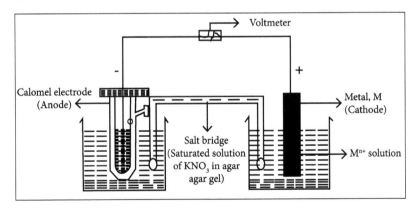

Galvanic cell obtained in this case is represented as, $Hg_{(l)} \mid Hg_2Cl_{2(s)} \mid KCl \parallel M^{+n} \mid M$

At anode, oxidation occurs, $2\,Hg + 2Cl^- \rightarrow Hg_2Cl_2 + 2e^-$

At cathode, reduction occurs, $M^{+n} + n^{e-} \rightarrow M$

EMF of cell, $E_{Cell} = E_{cathode} - E_{Anode}$

$$E_{Cell} = E_{Mn}+/M - ECE$$

$$E_{Mn}+/M = ECE + E_{cell}$$

3.2.2 Concentration Cells

Concentration cell is a type of galvanic cell in which electrode and electrolyte present in both the half cells are same but only the concentration of metal or electrolyte is different.

Two types of concentration cells:

*   Electrolyte concentration cell.

*   Electrode concentration cell.

Electrolyte Concentration Cells

It is a type of galvanic cell in which electrode & electrolyte present in both the half cells are same but only the concentration of electrolyte is different.

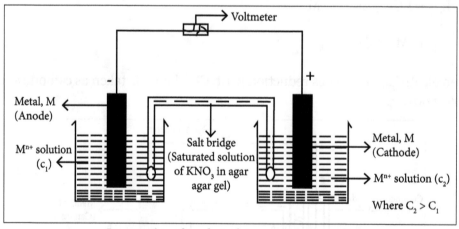

Construction of an electrolyte concentration cell.

Cell representation: $M / Mn + (C_1) || M n + (C_2) / M$

Metal immersed in the dilute solution will have lower potential act as anode,

$$M \rightarrow Mn + (C_1) + ne^-$$

Metal immersed in the concentrated solution will have higher potential act as cathode,

$$Mn + (C_2) + ne^- \rightarrow M$$

Net cell reaction is: $Mn + (C_2) \rightarrow Mn + (C_1)$

Nernst equation for electrolyte concentration cell is,

$$E_{Cell} = E^o_{Cell} + 0.0591 \log [Mn+] \text{cathode n} [Mn+] \text{Anode}$$

$$E^o_{Cell} = E^o_{cathode} - E^o_{anode} = 0$$

$$E_{Cell} = 0.0591 \log [Mn^+] \text{cathode n} [Mn^+] \text{Anode}$$

$$E_{Cell} = 0.0591 \log C_2 \text{ at 298k n} C_1$$

E_{Cell} is +ve and reaction is spontaneous only when $C_2 > C_1$

Problems

1. Let us write the electrode reactions and Calculate the EMF of the given cell at 298K, $Ag(s) AgNO_3$ (0.018M) $AgNO_3$ (1.2M)Ag(s).

Solution:

At anode: $Ag(s) \rightarrow Ag^+ + e^-$

At Cathode: $Ag^+ + e^- \rightarrow Ag(s)$

W.k.t $E_{cell} = \dfrac{0.0591}{n} \log \left(\dfrac{C_2}{C_1} \right)$ at 298K at 298K

$$E_{cell} = 0.0591 \log \left(\dfrac{1.2}{0.018} \right) (n = 1)$$

$$E_{cell} = 0.1078 \text{ V}$$

2. Let us calculate the emf of Copper concentration cell at 25° C, where the copper ions ratio in the cell is 10.

Solution:

$$\dfrac{[Cu^{+2}]_{cathode}}{[Cu^{+2}]_{anode}} = \dfrac{C_2}{C_1} = 10$$

W.k.t $E_{cell} = \dfrac{0.0591}{n} \log \left(\dfrac{C_2}{C_1} \right)$, at 298 K

$$E_{cell} = \frac{0.0591}{2} \log(10)$$

$$E_{cell} = 0.0296 \text{ V}$$

3. A cell contains two hydrogen electrodes. The negative electrode is in contact with a solution of 10^{-6} M hydrogen ions. The emf of the cell is 0.118volt at 25°C. Let us Calculate the concentration of hydrogen ions at the positive electrode.

Solution:

The cell may be represented as,

$$Pt \,|\, H_2 \,(1 \text{ atm}) \,|\, H^+ \,||\, H^+ \,|\, H_2 \,(1 \text{ atm}) \,|\, Pt$$

10^{-6} M CM

Anode: $H_2 \rightarrow 2H^+ + 2e^-$

Cathode: $2H^+ + 2e \rightarrow H^2$

$$E_{cell} = 0.0591/2 \log([H^+]_{cathode}2)/[10^{-6}]^2$$

$$0.081 = (0.0591) \log ([H^+])/10^{-6}$$

$$\log[H^+]_{cathode}/10^{-6} = 0.118/0.0591 = 2$$

$$[H^+]_{cathode}/10^{-6} = 10^2$$

$$[H^+]_{cathode} = 10^{-6} = 10^{-4} \text{ M}$$

4. The emf of the cell Ag|AgI in 0.05 MK\Sol. NH_4NO_3 |10.05 M $AgNO_3$\Ag is 0.788 volt at 25°C. The activity coefficient of KI and silver nitrate in the above solution is 0.90 each. Let us Calculate (i) the solubility product of AgI and (ii) the solubility of AgI in pure water at 25°C.

Solution:

Ag+ ion concentration on $AgNO_3$ side = 0.9 × 0.5 = 0.045 M

Similarly I- ion concentration in 0.05 M KI solution = 0.05 × 0.9 - 0.045 M

E_{cell} = 0.0591/1 log[Ag$^+$](R.H.S.)/[Ag$^+$](L.H.S.) = 0.0591 log 0.045/[Ag$^+$] (L.H.S.)

or

$\log 0.045/[Ag^+](L.H.S.) = 0.788/0.0591 = 13.33$

$[Ag^+]L.H.S. = 0.045/(2.138 \times 1013) = 2.105 \times 10^{-15}$ M

Solubility product of $AgI = \left[Ag^+\right]\left[I^-\right]$

$= 2.105 \times 10^{-15} \times 0.045$

$= 9.427 \times 10^{-17}$

Solubility of AgI = √(Solubility product of AgI)

$= \sqrt{\left(9.472 \times 10^{-17}\right)}$

$= 9.732 \times 10^{-9}$ g mol L^{-1}

$= 9.732 \times 10^{-9} \times 143.5$ g L^{-1}

$= 1.396 \times 10^{-6}$ g L^{-1}

3.3   Batteries: Dry Cell, Ni-Cd cells, Ni-Metal Hydride Cells, Li Cells and Zinc-air Cells

Battery

A battery is a device that consist of two or more galvanic cell connected in series or parallel or both, which converts chemical energy into electrical energy through redox reaction.

Example: Lead acid battery, Lithium ion battery, Nickel-Cadmium battery etc.

- Uses: Batteries are used in calculators, watches and pacemakers for heart, hearing aids, computers, car engines, emergency lightning in hospitals, military and space applications.

- Basic Components of Battery: The battery consists of four major components.

- Anode (Negative electrode): It releases electrons into the external circuit by undergoing oxidation,

$M \rightarrow Mn^+ + e^-$

- Cathode (Positive electrode): It accepts electrons coming from anode through external circuit,

$$Mn^+ + e^- \rightarrow M$$

- Electrolyte: It provides the medium for transfer of ions inside the cell between the anode and cathode. A solution of an acid, alkali or salt having high ionic conductivity is commonly used as an electrolyte.

- Separator: It is used to separate anode and cathode compartments in a battery to prevent internal short circuiting. It allows the ions from anode and cathode. Ex: Cellulose, nafion membranes.

Batteries

It is a device which converts the chemical energy into electrical energy and vice versa.

Ex: Lead-Acid battery and NICAD.

Emf Variation with Size

The terminal voltage of a battery cell depends on the chemicals and materials used in its construction and not on its physical size.

Batteries are classified into primary and secondary battery:

1. Primary Battery or Primary Cells

A battery which cannot be recharged, cell reactions are irreversible and discarded when the battery has delivered all its electrical energy.

Ex: dry cell or $Zn-MnO_2$ Cell, $Li-Mno_2$ cell

2. Secondary Battery

A battery which can be recharged, cell reactions are reversible and A battery which after discharging, can be recharged.

Ex: Lead storage cell and nickel cadmium cell.

3. Reserve Batteries

In reserve batteries, one of the components is stored separately and is incorporated into the battery when required.

Ex: Mg-AgCl and Mg-CuCl battery.

Characteristics of a Battery

- Voltage (cell potential).

- Current.

- Capacity.

- Electricity storage density.

- Energy Efficiency.

- Cycle Life.

- Self Life.

1. Voltage or cell potential (EMF): The voltage of a battery is given by the equation,

$$E_{cell} = (EC - EA) - \eta A - \eta C - iR_{cell}$$

Where EC and EA are the Electrode potentials of cathode and anode respectively, ηA and ηC are the over potentials at cathode and anode respectively and iRcell is the internal resistance of the cell.

Conditions to derive maximum voltage from a battery:

- Potential difference must be high.

- Over potentials at cathode and anode should be minimum.

- Internal resistance of the cell must be low.

2. Current: For efficient discharge, electron should flow at uniform rate in the electrolyte current is a measure of rate of flow of electrons during discharge. It is amount of charge flowing per unit and is expressed in ampere per second. Batteries provide direct current.

According to Ohms law, current $I = V/R$. At higher resistance more potential difference is required to force the current through the cell.

3. Capacity: Battery capacity is a measure of the charge stored by a battery, its corresponding SI unit is Ampere-hour (Ah) and is the amount of electrical energy the battery delivers over certain period. Capacity depends on the size of the battery.

The theoretical capacity may be calculated using Faradays relation,

$$C = wnF/M$$

Where,

C is the capacity

w is the mass of the active material.

M is its molar mass.

n is the number of moles electrons.

4. Electricity storage density: It's a measure of the charge per unit mass stored in the battery. The mass of battery includes masses of electrolyte current collectors, terminals and other subsidiary elements. Lighter subsidiary elements lead to high storage density. Example: 7g of lithium anode gives 96500 C whereas, for the same charge, 65g of zinc would be required.

5. Energy efficiency: It is defined as the ratio of useful energy output to the total energy input. A battery should have high energy efficiency.

$$\% \text{ Energy efficiency} = \frac{\text{Energy released during discharge} \times 100}{\text{Energy required during recharge}}$$

6. Cycle-life: Cycle life of a battery is the number of charge-discharge cycle that can be achieved before failure occurs. The greater is the average depth of discharge, the shorter the cycle life. It is applicable only for the secondary batteries.

It is necessary that during charging the active material is regenerated in a suitable state for discharge. The discharge-charge cycle depends on chemical composition and distribution of active materials in the cell.

7. Shelf-life: Some of the batteries can be stored for many years. The duration of storage of a cell without self-discharge or loss of performance is called shelf-life.

Dry-Cell

Recently the most popular dry-cell battery to be used has been the alkaline-cell battery. In the zinc-carbon battery, the zinc is not easily dissolved in basic solutions. Though it is fairly cheap to construct a zinc-carbon battery, the alkaline-cell battery is favored because it can last much longer.

Instead of using NH4Cl as an electrolyte the alkaline-cell battery will use NaOH or KOH instead. The same reaction will occur where zinc is oxidized and it will react with OH− instead.

$$Zn^{2+} + 2OH^- \rightarrow Zn(OH)^2$$

Once the chemicals in the dry-cell battery can no longer react together, the dry-cell battery is dead and cannot be recharged. Alkaline electrochemical cells have a much longer lifetime but the zinc case still becomes porous as the cell is discharged and the substances inside the cell are still corrosive. Alkaline cells produce 1.54 volts.

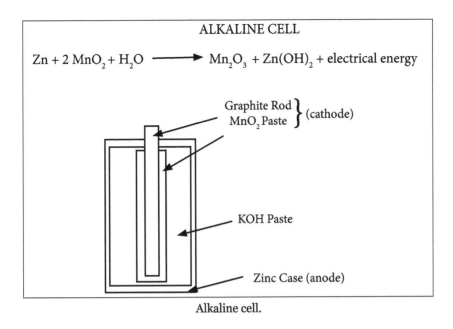

ALKALINE CELL

$$Zn + 2\,MnO_2 + H_2O \longrightarrow Mn_2O_3 + Zn(OH)_2 + \text{electrical energy}$$

Graphite Rod
MnO_2 Paste $\Big\}$ (cathode)

KOH Paste

Zinc Case (anode)

Alkaline cell.

Nickel–cadmium Battery (Ni-Cd Cells)

A nickel - cadmium storage cell consists of cadmium as anode and NiO2paste as cathode and KOH as the electrolyte.

The cell is represented as $Cd \mid Cd(OH)_2 \,\|KOH(aq) \mid NiO_2 \mid Ni$.

Construction and Working

When the nickel battery operates, Cd is oxidized to Cd^{2+} ions at anode and the insoluble $Cd(OH)_2$ is formed. NiO is reduced to Ni^{2+} ions which further combines with OH- ions to form $Ni(OH)_2$. It produces about 1.4 V. The following cell reactions occur.

Anodic reaction: $Cd(s)+2OH^- \rightarrow Cd(OH)2(s)+2e^-$

Cathodic reaction: $NiO_2(s)+2H_2O+ 2e^- \rightarrow Ni(OH)_2(s)+2OH^-$

Overall cell reaction during discharging,

$$Cd(s)+ NiO_2(s)+2H_2O \rightarrow Cd(OH)_2(s)+Ni(OH)_2(s)+Energy$$

From the above cell reactions, it is clear that $Cd(OH)_2$ and $Ni(OH)_2$ are deposited at both the anodes and cathodes respectively. So this can be reversed by recharging the cell.

Overall Cell Reaction During Recharging

The cell can be charged by passing electric current in the opposite direction. The electrode reactions get reversed. As a result Cd is deposited on the anode and NiO_2 on the cathode.

$$Cd(OH)_2(s) + Ni(OH)_2(s) + Energy \underset{\text{discharging}}{\overset{\text{charging}}{\rightleftharpoons}} Cd(s) + NiO_2(s) + 2H_2O.$$

Advantages of Ni-Cd Battery

- It is portable and rechargeable cell.
- It has longer life than lead - acid battery.
- It can be easily packed like dry cell since it is smaller and lighter.

Uses

- It is used in calculators.
- It is used in gas electronics flash units.
- It is used in transistors, cordless electronic appliances, etc.

Nickel-Metal Hydride Cell

- Anode — A metal hydride, MH.
- Cathode — Nickel oxy hydroxide/Ni.
- Electrolyte — KOH.

A relatively new technology is adopted in the case of the chargeable sealed nickel-metal hydride battery with characteristics similar to those of the sealed Ni-Cd batteries. The Ni-MH battery uses hydrogen absorbed in a metal alloy for the active negative material whereas cadmium is used in the Ni-Cd battery and that makes the noticeable difference between the two.

A higher energy density can be achieved in the case of metal hydride electrode than the cadmium electrode. Thus, a smaller amount of the negative electrode is used in the Ni-metal hydride. This allows for a larger volume for the +ve electrode, which results in a higher capacity or longer service life for the metal hydride battery.

Moreover, as the metal hydride battery is free of Cd, it is considered more environmentally friendly than the Ni-Cd battery and may reduce the problems associated with the disposal of rechargeable nickel batteries.

Nickel-metal hydride batteries consist of a positive plate of a highly porous sintered or felt nickel substrate impregnated with nickel hydroxide as its principal active material, a negative plate of a highly porous structure using a perforated $LaNi_5$ alloy grid (a hydrogen-absorbing alloy). A synthetic non-woven material separates the two electrodes, which serves as a medium for absorbing the electrolyte and a sealing plate provided with a self-resealing safety vent.

The chemistry of the electrode reactions of Ni-MH battery can be described as follows:

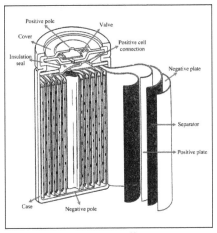

Ni-MH cell.

In the charged state of Ni-MH battery, nickel oxy-hydroxide is the active material of the +ve electrode. This is same as the positive electrode in the Ni-Cd battery. In the charged state of the Ni-MH battery, hydrogen is stored in a hydrogen absorbing alloy as metal hydride, $LaNi_5$, (-ve active material).

This metal alloy is capable of undergoing a reversible hydrogen absorbing-desorbing reaction as the battery is charged and discharged. An aqueous solution of KOH is the major component of the electrolyte, with minimum amount of the electrolyte absorbed by the separator and the electrodes.

As can be seen from the overall reaction given below, the chief characteristics of the principle behind a Ni-MH battery is that hydrogen moves from the +ve to -ve electrode during charge and in reverse order during the discharge with the electrolyte taking no part in the reaction, which means that there is no accompanying increase or decrease in electrolyte.

The discharge electrode reactions of the Ni-MH battery are as follows:

At anode,

$$MH + OH^- \leftrightarrow M + H_2O + e; \ E^\circ = 0.83 \ V$$

At cathode,

The nickel oxyhydroxide is reduced to nickel hydroxide.

$$NiOOH + H_2O + e \leftrightarrow Ni(OH)_2 + OH-; \ E^\circ = 0.52 \ V$$

The overall reaction on discharge is,

$$MH + NiOOH \leftrightarrow M + Ni(OH)_2 \; ; \; E° \; 1.35 \; V$$

The process is reversed during charging of the Ni-MH battery.

Advantages

The following are the advantages of an Ni-MH battery:

- High capacity.

- No maintenance required.

- Minimum environmental problem.

- Rapid recharging capability.

- Long cycle life.

- Long shelf life in state of charge.

Applications

Nickel-metal hydride batteries are used in computers, cellular phones and other portable and consumer electronic applications where high specific energy is required.

Lithium cells: Introduction, construction, working and applications of Li-MnO$_2$ and Li-ion batteries.

Lithium is a theoretically 'active material for negative electrode' of the electrochemical cells owing to its least noble nature and low specific gravity.

- 'Primary cells' with metallic lithium electrodes and non-aqueous electrolytes were successfully introduced into the market. The outstanding features in comparison with conventional batteries with aqueous electrolytes are—high voltage, high energy density [both volumetric and gravimetric energy densities are high], low self-discharging rate and, wide range of operation.

- Thus, 'secondary batteries' with metallic lithium negative electrodes have attracted much attention as a candidate for the battery with high energy density and much effort has been made in developing secondary lithium batteries.

Many practical problems, however, have been encountered in the development of rechargeable lithium batteries. Among them are:

- Poor cycle life.

- Need for long charging time.

- Poor safety characteristics.

Li-ion Battery: Secondary Battery

Construction

- Anode: Lithium atoms intercalated in layered graphite / Carbon metal.

- Cathode: Lithium atoms intercalated in layered structure of MO_2.

- Electrolyte: The electrolyte is lithium salt such as $LiPF_6$ dissolved in organic solvents (Ethylene carbonate – dimethyl carbonate).

- Separator: Made of Poly propylene membrane.

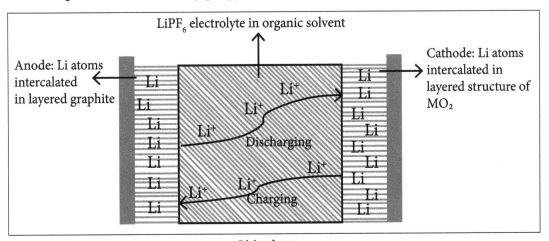

Li-ion battery.

Cell representation: $Li \mid Li^+, C/LiPF_6$ in organic solvents $\mid Li - MO_2$.

Electrode reactions are:

During discharging, lithium atoms present in graphite layer are oxidized, liberating electrons and lithium ions. Electrons flows through external circuit to cathode and lithium ions flow through the organic electrolyte towards cathode.

At anode : $Li - C_6 \rightarrow Li^+ + e^- + 6C$

At cathode, lithium ions are reduced to lithium atoms and are inserted in to the layered structure of metal oxide.

At cathode : $MO_2 + Li^+ + e^- \rightarrow Li - MO_2$

During charging, lithium atoms present in layered structure of metal oxide are oxidized, liberating electrons and lithium ions. Electrons flows through external circuit and lithium ions flow through the organic electrolyte towards graphite carbon electrode.

$$\text{At anode}: \text{Li}-\text{MO}_2 \rightarrow \text{MO}_2 + \text{Li}^+ + e^-$$

At graphite electrode, lithium ions are reduced to lithium atoms and are inserted in to the layered structure of graphite.

$$\text{At cathode}: \text{Li}^+ + e^- + 6\text{C} \rightarrow \text{Li}-\text{C}_6$$

Applications: Used in cell phone, Laptops, Electrical vehicles, Aerospace applications, Portable LCD TV etc.

Zinc-Air Battery

- Anode — Granulated Zn powder.

- Cathode — Air/C.

- Electrolyte — KOH 6M.

The production of electrochemical energy in Zn/air battery is due to the use of oxygen from the atmosphere. The diffused oxygen acts as a cathode reactant in the battery. The air cathode catalytically promotes the reaction of oxygen with an aqueous alkaline electrolyte and is not consumed or changed during the discharge.

When an alkaline electrolyte is used in the Zn/air battery, it is necessary to increase only the amount of zinc present to increase battery capacity.

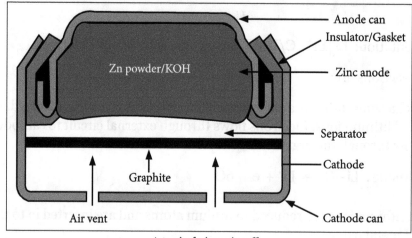

A typical zinc-air cell.

The air cathode acts only as a reaction site and is not consumed. The reason for the increased energy density in the Zn/air cell is because of the larger volume containing the active material. Since the air cathode has infinite life, the electrical capacity of the cell is determined only by the anode capacity, resulting in at least a doubling of energy density.

Theoretically, the air cathode has infinite use life and its physical size and its electrochemical properties remain unchanged during cell discharge. A schematic representation of a typical Zn/air cell is shown in the above figure.

A loose granulated powder of Zn mixed with electrolyte [KOH] acts as the zinc anode material and in some cases, a gelling agent is used to mobilize the composite and ensure adequate electrolyte contact with zinc granules. The outer metal (button type) acts as the cathode of the battery and a plastic gasket insulates the anode active materials and the cathode as shown in the above figure.

The chemistry of the electrode reactions taking place in zinc/air battery are as follows:

At anode,

$$Zn \rightarrow Zn^{2+} + 2e; \; E^{\circ} = 1.20 \text{ V}$$
$$Zn^{2+} + 2OH^{-} \rightarrow Zn(OH)2$$
$$Zn(OH)_2 \rightarrow ZNO + H_2O$$

At cathode,

$$\tfrac{1}{2} O_2 + H_2O + 2e \rightarrow 2OH^{-}; \; E^{\circ} = 0.45 \text{ V}$$

The overall efficiency of zinc/air battery,

$$Zn + \tfrac{1}{2} O_2 \leftrightarrow ZnO; \; E^{\circ} = 1.65 \text{ V}$$

The output of the zinc-air battery is 1.65 V.

Advantages

Zn/air battery technology offers the following advantages for many applications:

- High energy density.
- Flat discharge voltage.
- Long shelf life.
- No ecological problems.
- Low cost.
- Capacity independent of load and temperature.

Applications

Zn-air batteries have been most successfully employed as a power source for hearing aids. Other applications include electronic pagers, voice transmitters, portable battery chargers, various medical devices and so on.

3.4   Corrosion and its Theories

Corrosion is the deterioration of a metal as a result of chemical reactions between it and the surrounding environment.

$$\text{Metal} \xleftrightarrow[\text{Extraction}]{\text{Corrosion}} \text{Metal ore}$$

Both the type of metal and the environmental conditions, particularly what gases that are in contact with the metal, determine the form and rate of deterioration.

Corrosion is an undesirable process. Due to corrosion there is limitation of progress in many areas. The cost of replacement of materials and equipments lost through corrosion is unlimited.

Metals and alloys are used as fabrication or construction materials in engineering. If the metals or alloy structures are not properly maintained, they deteriorate slowly by the action of atmospheric gases, moisture and other chemicals. This phenomenon of destruction of metals and alloys is known as corrosion.

Corrosion of metals is defined as the spontaneous destruction of metals in the course of their chemical, electrochemical or biochemical interactions with the environment. Thus, it is exactly the reverse of extraction of metals from ores.

Example: Rusting of iron.

A layer of reddish scale and powder of oxide $\left(Fe_3O_4\right)$ is formed on the surface of iron metal.

A green film of basic carbonate $\left[CuCO_3 + Cu(OH)_2\right]$ is formed on the surface of copper, when it is exposed to moist-air containing carbon dioxide.

Consequences

- Efficiency of the machine is lost due to corrosion products.
- Products get contaminated due to released toxic products.
- Corroded equipment must be replaced frequently.
- Failure of plants.

- Necessity of over designing.

Causes of Corrosion

In nature, metals occur in two different forms:

- Native state.

- Combined state.

Native State

The metals exist as such in the earth crust then the metals are present in a native state.

Native state means free or uncombined state. These metals are non-reactive in nature. They are noble metals which have very good corrosion resistance. Example: Au, Pt, Ag, etc.

Combined State

Except noble metals, all other metals are highly reactive in nature which undergoes reaction with their environment to form stable compounds called ores and minerals. This is the combined state of metals.

Example: Fe_2O_3, ZnO, PbS, C_aCO_3 etc.

Metallic Corrosion

The metals are extracted from their metallic compounds (ores). During the extraction, ores are reduced to their metallic states by applying energy in the form of various processes.

In the pure metallic state, the metals are unstable as they are considered in excited state (higher energy state). Therefore as soon as the metals are extracted from their ores the reverse process begins and form metallic compounds which are thermodynamically stable (lower energy state).

Hence, when metals are used in various forms, they are exposed to environment and when heated, the exposed metal surface begin to decay (conversion to more stable compound). This is the basic reason for metallic corrosion.

$$\text{Metal} \underset{\text{Metallurgy}-\text{Reduction}}{\overset{\text{Corrosion}-\text{Oxidation}}{\rightleftarrows}} \text{Metallic Compound} + \text{Energy}$$

Although corroded metal is thermodynamically more stable than pure metal but due to corrosion, useful properties of a metal like malleability, ductility, hardness, luster and electrical conductivity are lost.

Effects of Corrosion

The economic and social consequences of corrosion includes:

- Due to formation of corrosion product over the machinery, the efficiency of the machine gets failed and leads to plant shut down.

- The products contamination or loss of products occurs due to corrosion.

- The corroded equipment must be replaced.

- Preventive maintenance like metallic coating or organic coating is required.

- Corrosion releases the toxic products.

- Health (e. g., from pollution due to a corrosion product or due to the escaping chemical from a corroded equipment).

Classification or Theories of Corrosion

Based on the environment, corrosion is classified into:

- Dry or chemical corrosion.
- Wet or electrochemical corrosion.

1. Dry or Chemical Corrosion

This type of corrosion is due to the direct chemical attack on metal surfaces by the atmospheric gases such as oxygen, halogen, hydrogen sulphide, sulphur dioxide, nitrogen or anhydrous inorganic liquid, etc. The chemical corrosion is defined as the direct chemical attack on metals by the atmospheric gases present in the environment.

Example

- Silver materials undergo chemical corrosion by Lower case atmospheric H_2S gas.

- Iron metals undergo chemical corrosion by HCL gas.

Types of Dry or Chemical Corrosion

- Corrosion by oxygen or oxidation corrosion.
- Corrosion by hydrogen.
- Liquid metal corrosion.

Corrosion by Oxygen or Oxidation Corrosion

Oxidation corrosion is brought about by the direct attack of oxygen at low or high temperature on metal surfaces in the absence of moisture. Alkali metals (Li, Na, K etc.,) and alkaline earth metals (Mg, Ca, Sn, etc.,) are rapidly oxidized at low temperature. At high temperature, almost all metals (except Ag, Au and Pt) are oxidized. The reactions of oxidation corrosion are as follows.

Mechanism

1. Oxidation takes place at the surface of the metal forming metal ions.

$$M \rightarrow M_2^{+} + 2e^{-}$$

2. Oxygen is converted to oxide ion $(O_2 -)$ due to the transfer of electrons from metal.

$$n/2 \, O_2 + 2ne^{-} \rightarrow n \, O_2^{-}$$

3. The overall reaction is of oxide ion reacts with the metal ions to form metal oxide film.

$$2 \, M + n/2 \, O_2 \rightarrow 2 \, Mn^{+} + nO_2^{-}$$

The Lower case Nature of the Lower case Oxide formed plays an important part in oxidation corrosion process.

$$\text{Metal + Oxygen} \rightarrow \text{Metal oxide (corrosion product)}$$

When oxidation starts, a thin layer of oxide is formed on the metal surface and the nature of this film decides the further action.

i. Stable Layer

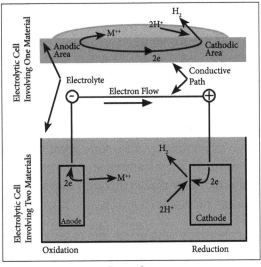

Corrosion

A stable layer is fine grained in structure and can get adhered tightly to the parent metal surface. Hence, such layer can be of impervious nature (i.e., which cuts-off penetration of attaching oxygen to the underlying metal).

Such a film behaves as protective coating in nature, thereby shielding the metal surface. The oxide films on Al, Sn, Pb, Cu, Pt, etc., are stable, tightly adhering and impervious in nature.

ii. Unstable Oxide Layer

This is formed on the surface of noble metals such as Ag, Au, Pt. As the metallic state is more stable than oxide, it decomposes back into the metal and oxygen. Hence, oxidation corrosion is not possible with noble metals.

iii. Volatile Oxide Layer

The oxide layer film volatilizes as soon as it is formed. Hence always a fresh metal surface is available for further attack. This causes continuous corrosion. MoO_3 is volatile in nature.

iv. Porous Layer

The layer having pores or cracks. In such a case, the atmospheric oxygen have access to the underlying surface of metal through the pores or cracks of the layer, thereby the corrosion continues unobstructed, till the entire metal is completely converted into its oxide.

Pilling-Bedworth Rule

According to it "an oxide is protective or non-porous, if the volume of the oxide is as great as the volume of the metal from which it is formed". On the other hand, "if the volume of the oxide is less than the volume of metal, the oxide layer is porous (or non-continuous) and hence non-protective because it cannot prevent the access of oxygen to the fresh metal surface below".

Thus alkali and alkaline earth metals like (Li, K, Na, Mg) form oxides of volume less than the volume of metals. Consequently, the oxide layer faces stress and strains thereby developing cracks and pores in its structure. Porous oxide scale permits free access of oxygen to the underlying metal surface (through cracks and pores) for fresh action and thus, corrosion continues non-stop.

Metals like Aluminium forms oxide whose volume is greater than the volume of metal. Consequently an extremely tightly-adhering non-porous layer is formed. Due to the absence of any pores or cracks in the oxide film, the rate of oxidation rapidly decreases to zero.

Corrosion by other Gases (by Hydrogen)

a. Hydrogen Embrittlement

Loss in ductility of a material in the presence of hydrogen is known as hydrogen embrittlement.

Mechanism

This type of corrosion occurs when a metal is exposed to hydrogen environment. Iron liberates atomic hydrogen with hydrogen sulphide in the following way,

$$Fe + H_2S \rightarrow FeS + 2H$$

Hydrogen diffuses into the metal matrix in this atomic form and gets collected in the voids present inside the metal. Further, diffusion of atomic hydrogen makes them combine with each other and forms hydrogen gas,

$$H + H \rightarrow H_2 \uparrow$$

Collection of these gases in the voids develops very high pressure, causing cracking or blistering of metal.

b. Decarburization

The presence of carbon in steel gives sufficient strength to it. But when steel is exposed to hydrogen environment at high temperature, atomic hydrogen is formed,

$$H_2 \xrightarrow{\text{Heat}} 2H$$

Atomic hydrogen reacts with the carbon of the steel and produces methane gas,

$$C + 4H \rightarrow CH_4$$

Hence the carbon content in steel is decreases. The process of decrease in carbon content in steel is known as decarburization. Collection of methane gas in the voids of steel develops high pressure which causes cracking. Thus steel loses its strength.

Liquid Metal Corrosion

This is due to chemical action of flowing liquid metal at high temperatures on solid metal or alloy. Such corrosion occur in devices used for nuclear power. The corrosion reaction involves either,

- Dissolution of a solid metal by a liquid metal.

Or

- Internal penetration of the liquid metal into the solid metal.

Both these modes of corrosion cause weakening of the solid metal.

2. Wet or Electrochemical Corrosion

Electrochemical corrosion involves:

- The formation of anodic and cathodic areas or parts in contact with each other.

- Presence of a conducting medium.

- Corrosion of anodic areas.

- Formation of corrosion product somewhere between anodic and cathodic areas.

This involves flow of electron-current between the anodic and cathodic areas.

At anodic area, oxidation reaction takes place (liberation of free electron), so anodic metal is destroyed by either dissolving or assuming combined state (such as oxide, etc). Hence corrosion always occurs at anodic areas.

$$M_{(metal)} \rightarrow Mn^+ + n\ e^-$$

Mn(metal ion) → Dissolves in solution → forms compounds such as oxide

At cathodic area, reduction reaction takes place (gain of electrons), usually cathode reactions do not affect the cathode since most metals cannot be further reduced. So at cathodic part dissolved constituents in the conducting medium accepts the electrons to form some ions like OH and O_2^-.

Cathodic reaction consumes electrons either by:

- Evolution of hydrogen.

- Absorption of oxygen depending on the nature of the corrosive environment.

Hydrogen Evolution Type

All metals above hydrogen in the electrochemical series have a tendency to get dissolved in acidic solution with simultaneous evolution of hydrogen.

It occurs in acidic environment. Consider the example of iron.

At anode : $Fe \rightarrow Fe^{2+} + 2e^-$

These electrons flow through the metal from anode to cathode where H+ ions of acidic solution are eliminated as hydrogen gas.

At cathode : $2\,H^+ + 2\,e^- \rightarrow H_2 \uparrow$

The overall reaction is: $Fe + 2H^+ \rightarrow Fe_2{}^+ + H_2$

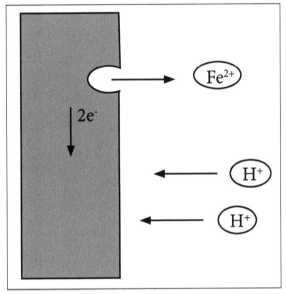

Hydrogen evolution type.

Oxygen Absorption Type

Rusting of iron in neutral aqueous solution of electrolytes like (NaCl solution) in the presence of atmospheric oxygen is a common example of this type of corrosion. The surface of iron is usually coated with a thin film of iron oxide.

However, if this iron oxide film develops some cracks, anodic areas are created on the surface, while the well metal parts acts as cathodes.

At Anode: Metal dissolves as ferrous ions with liberation of electrons,

$Fe \rightarrow Fe^{2+} + 2e^-$

Cathode: The liberated electrons are intercepted by the dissolved oxygen,

$\tfrac{1}{2}\,O_2 + H_2O + 2\,e^- \rightarrow 2OH$

The Fe2+ ions and OH- ions diffuse and when they meet ferrous hydroxide is precipitated,

$Fe^{2+} + 2OH^- \rightarrow Fe(OH)_2$

(i) If enough oxygen is present, ferrous hydroxide is easily oxidized to ferric hydroxide,

$$4Fe(OH)_2 + O_2 + 2H_2O \rightarrow 4Fe(OH)_3 \quad (\text{Yellow rust } Fe_2O_3 \cdot H_2O)$$

(ii) If the supply of oxygen is limited, the corrosion product may be even black anhydrous magnetite Fe_3O_4.

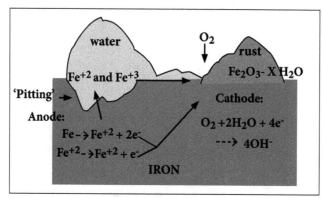

Oxygen absorption type.

Types of Electrochemical Corrosion

The electrochemical corrosion is classified into the following two types:

- Galvanic corrosion.

- Differential aeration or concentration cell corrosion.

1. Galvanic Corrosion

When two dissimilar metals (e.g., zinc and copper) are electrically connected and exposed to an electrolyte, the metal higher in electrochemical series undergoes corrosion. In this process the more active metal (with more negative electrode potential) acts as a anode while the less active metal (with less negative electrode potential) acts as cathode.

In the above example, zinc (higher in electrochemical series) forms the anode and is attacked and gets dissolved whereas copper (lower in electrochemical series or more noble) acts as cathode.

Mechanism

In acidic solution the corrosion occurs by the hydrogen evolution process while in neutral or slightly alkaline solution, oxygen absorption occurs. The electron-current flows from the anode metal, zinc to the cathode metal, copper.

$$Zn \rightarrow Zn^{2+} + 2e^- \ (Oxidation)$$

Thus it is evident that the corrosion occurs at the anode metal, while the cathodic part is protected from the attack.

Example:

- Steel screws in a brass marine hardware.

- Lead-antimony solder around copper wire.

- A steel propeller shaft in bronze bearing.

- Steel pipe connected to copper plumbing.

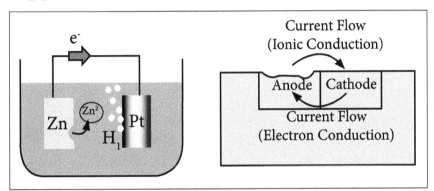

Galvanic corrosion.

2. Concentration Cell Corrosion

It is due to electrochemical attack on the metal surface, exposed to an electrolyte of varying concentrations or of varying aeration. It occurs when one part of metal is exposed to a different air concentration from the other part. This causes a difference in potential between differently aerated areas. It has been found experimentally that poor-oxygenated parts are anodic.

Examples:

- The metal part immersed in water or in a conducting liquid is called water line corrosion.

- The metal part is partially buried in soil.

Explanation

If a metal is partially immersed in a conducting solution the metal part above the solution is more aerated and becomes cathodic. The metal part inside the solution is less aerated and thus becomes anodic and suffers corrosion.

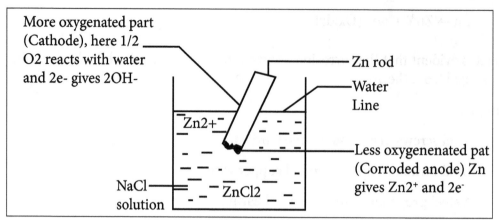

At anode: Corrosion occurs (less aerated) $M \rightarrow M^{2+} + 2e^-$

At Cathode: OH^- ions are produced (more aerated).

$$\tfrac{1}{2} O_2 + H_2O + 2e^- \rightarrow 2OH^-$$

Examples for this type of corrosion are:

- Pitting or localized corrosion.

- Crevice corrosion.

- Pipeline corrosion.

- Corrosion on wire fence.

Pitting Corrosion

Pitting is a localized attack which results in the formation of a hole around which the metal is relatively un attacked. The mechanism of this corrosion involves setting up of differential aeration or concentration cell. Metal area covered by a drop of water, dust, sand, scale etc. is the aeration or concentration cell.

Pitting corrosion is explained by considering a drop of water or brine solution on a metal surface. The area covered by the drop of salt solution has less oxygen and acts as anode. This area suffers corrosion, the uncovered area acts as cathode due to high oxygen content.

It has been found that the rate of corrosion will be more when the area of cathode is larger and the area of the anode is smaller. Hence there is more material around the small anodic area results in the formation hole or pit.

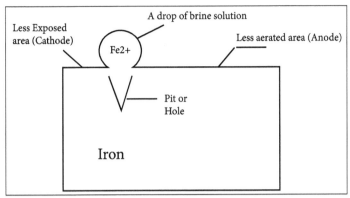

Pitting corrosion.

At anode: Fe is oxidized to Fe^{2+} and releases electrons,

$$Fe^- \rightarrow Fe^{2+} + 2e^-$$

At cathode: Oxygen is converted to hydroxide ion,

$$\frac{1}{2} O_2 + H_2O + 2e^- \rightarrow 2OH^-$$

The net reaction is $Fe + 2OH^- \rightarrow Fe(OH)_2$

The above mechanisms can be confirmed by using ferroxyl indicator (a mixture containing phenolphthalein and potassium ferricyanide). Since OH^- ions are formed at the cathode, this area imparts pink colour with phenolphthalein indicator. At the anode, iron is oxidized to Fe^{2+} which combines with ferricyanide and shows blue colour.

Crevice Corrosion

If a crevice (a crack forming a narrow opening) between metallic and non-metallic material is in contact with a liquid, the crevice becomes anodic region and undergoes corrosion. Hence oxygen supply to the crevice is less. The exposed area has high oxygen supply and acts as cathode.

Bolts, nuts, rivets, joints are examples for this type of corrosion.

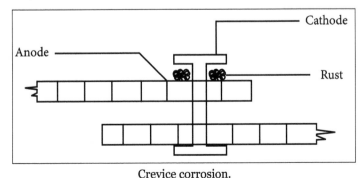

Crevice corrosion.

Pipeline Corrosion

Buried pipelines or cables passing from one type of soil (clay, less aerated) to another soil (sand, more aerated) may get corroded due to differential aeration.

Corrosion in Wire Fence

A wire fence is one in which the areas where the wires cross (anodic) are less aerated than the rest of the fence (cathodic). Hence corrosion takes place at the wire crossing.

Corrosion occurring under metal washers and lead pipeline passing through clay to cinders (ash) are other examples.

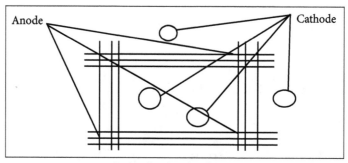

Corrosion in wire fence.

3.5 Formation of Galvanic Cells

Formation of Galvanic Cells by Different Metals

It occurs when two dissimilar metals (galvanic couples) are in contact with each other in a corrosive conductive medium, a potential difference is set up resulting in a galvanic current. The two metals differ in their tendencies to undergo oxidation.

The metal with lower electrode potential or more active metal acts as the anode and metal with higher electrode potential acts as the cathode. The potential difference is main factor for corrosion to take place. The anodic metal undergoes corrosion where as cathodic metal gets unaffected.

Eg: When iron is in contact with copper, iron has lower electrode potential acts as anode and undergo oxidation as,

$$Fe \rightarrow Fe^{2+} + 2e^-$$

Copper having higher electrode potential acts as cathode is unaffected. Depending on the corrosive environment near cathode either hydrogen evolved or oxygen absorbed resulting in hydroxide ion formation.

At cathode,

$$O_2 + 2H_2O + 4e^- \rightarrow 4\,OH^-$$

$$2Fe^{2+} + 4OH^- \rightarrow 2Fe(OH)_2$$

$4Fe(OH)_2 + O_2 + 2H_2O \rightarrow 2\left[Fe_2O_3 . 3H_2O\right]$. Yellow rust (in limited oxygen black rust formed).

The Rate of Galvanic Corrosion Depends Upon

Potential difference between anodic and cathodic metals, ratio of the anodic and cathodic area and environmental factors and tendency of the metal to undergo passivity etc.

Other e.g: When Fe is in contact with Sn then Fe acts as anode and Sn acts as cathode but when Fe contact with the Zn, Fe acts as cathode where as Zn acts as anode.

Formation of Galvanic Cells by Concentration Cells

A concentration cell is an electrochemical cell in which both half-cells are of the same type, but with different electrolyte concentrations.

The following cell notations are examples of concentration cells:

$$Cu \mid Cu^{2+}(aq,\ 0.0010\ M) \parallel Cu^{2+}(aq,\ 1.0\ M) \mid Cu$$

$$Ag \mid Ag^+(aq, 0.0010\ M) \parallel Ag^+(aq,\ 0.10\ M) \mid Ag$$

In concentration cells, the half-cell with the lower electrolyte concentration serves as an anode half-cell and one with the higher electrolyte concentration is the cathode half-cell. At the anode half-cell, the oxidation reaction occurs to increase the electrolyte concentration and at the cathode half-cell, a reduction reaction occurs to decrease its electrolyte concentration.

Oxidation-reduction reaction will continue until electrolyte concentrations in both half-cells become equal.

At anode half-cell: $Cu(s) \rightarrow Cu^{2+}(aq) + 2e^-$; $\left(in\ 0.0010\ M\ Cu^{2+}\right)$

At cathode half-cell: $Cu^{2+}(aq) + 2e^- \rightarrow Cu(s)$; $\left(in\ 0.50\ M\ Cu^{2+}\right)$

Formation of Galvanic Cells by Differential Aeration Corrosion

The differential aeration corrosion occurs when metal surface is exposed to the differential air or oxygen concentration. This develops galvanic cell (O_2 concentration cell) on metal and initiates the corrosion.

The part of the metal exposed to lower oxygen concentration acts as anode and the part of the metal exposed to higher concentration acts as cathode, so that poorly oxygenated region undergoes corrosion.

Ex: when a metal strip of iron is partially immersed in an aerated solution of NaCl the concentration of O_2 is higher at the top surface than inside the solution. Since cathodic reaction requires oxygen, the cathodic area tends to concentrate near the water line so that bottom portion of specimen acts as anode where corrosion starts.

At anode: $Fe \rightarrow Fe^{2+} + 2e^-$

At cathode (near water line): $O_2 + 2H_2O + 4e^- \rightarrow 4OH^-$

There are two types:

- Water line corrosion.

- Pitting corrosion.

Formation of Galvanic Cells by Waterline Corrosion

This is a case of differential aeration corrosion commonly observed in steel water tanks, ocean going ships etc. in which portion of the metal is always under water.

The part of the metal below water line is exposed only to dissolved oxygen while the part above water line is exposed to higher concentration of atmospheric oxygen.

Thus the metal part below water line acts as anode and undergoes corrosion. Whereas the metal part above waterline, which is more oxygenated acts as cathode and unaffected. This type of corrosion is commonly observed in the ships floating in seawater for a long period of time.

3.6 Passivity of Metals, Pitting Corrosion and Galvanic Series

Passivity of Metals

Passivity is defined as a condition of corrosion resistance due to formation of thin surface film under oxidizing conditions, some metals and alloys having simple barrier films with reduced corrosion of active potential.

For example in the presence of concentrated "fuming" nitric acid, iron is virtually inert despite the highly oxidizing conditions of the solution when the acid is diluted with water the iron remains inert, while initially corrodes vigorously evolving brown, nitrous oxide gas when the surface is lightly scratched. Passivity, is displayed by the chromium, Aluminum, iron (in some environment) nickel, titanium and many of their alloys.

It is felt that this passive behavior results from formation of a highly adherent and very thin oxide film on the metal surface, which serves as a protective barrier to further corrosion. Stainless steels (iron alloy) are highly resistant to corrosion in a rather wide variety of the atmospheres as a result of passivation.

Chromium is noted for formation of very stable, thin resistant surface in less oxidizing conditions when alloys with other metals especially iron as shown in stainless steel which is having minimum 12% Cr and it is passive in most aerated solutions.

Pitting Corrosion

It is a localized and accelerated corrosion. Pitting corrosion is an autocatalytic process, When a small particles of dust or water etc are get deposited on a metal (like steel). The portion covered by the dust will not be well-aerated area compared to the exposed surface hence the covered surface becomes anodic with respect to the surface exposed.

In presence of a conducting medium (moisture) corrosion starts below dust part forming a pit. Once pit is formed the ratio of corrosion increases, because of formation of smaller anodic and larger cathodic area intense corrosion takes place. The pit grows and ultimately may cause failure of the metal.

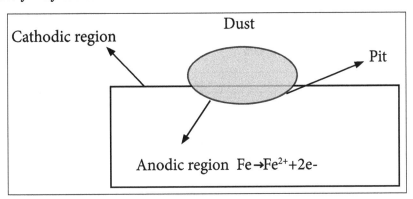

At anode: $Fe \rightarrow Fe^{2+} + 2e^-$

At cathode: $O_2 + 2H_2O + 4e^- \rightarrow 4OH^-$

Galvanic Series

An arrangement of metals and alloys in the order of their corrosion resistance in given environment is known as Galvanic Series.

Different metals have different tendencies to react or corrode. The familiar electro-chemical series lists various elements in the order of their standard electrode potentials the electrochemical series has however, limited utility in corrosion studies and a poor guide in predicting corrosion behavior of metals and alloys in real environment since it does not take passivity of the metal into the account.

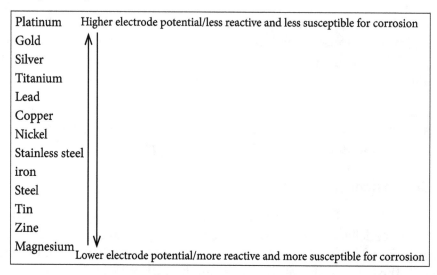

Eg: Chromium is less noble than iron in ECS, But Cr is a passive metal & its alloys with iron import excellent corrosion resistance. This is contrary to the prediction by ECS.

In order to overcome these limitations galvanic series was introduced. Corrosion studies on various metals and alloys were actually performed in various environments. Then metals and the alloys were arranged in the order of their corrosion resistance.

Difference between ECS & GS

Electrochemical series	Galvanic Series
1. E° values are measured by dipping metal in their salt solution.	Developed by studying the corrosion of metal or alloy in the given environment.
2. Position of the given metal is fixed.	The Position may shift.
3. No information regarding position of the alloys.	The alloys are included in the series.
4. It predicts relative displacement tendency.	It predicts relative corrosion tendency.

3.6.1 Factors Which Influence the Rate of Corrosion

The factors that affect the rate of corrosion are:

1. Factors Connected with the Metal

- Position of the Metal in the EMF Series: The type of impurity present in it and its electro-positive nature decides the corrosion of a metal. For example when iron has impurities like copper, tin, etc., iron corrodes since iron is more electro-positive than metals like copper and tin. On the other hand when the iron is coupled with zinc, zinc corrodes since zinc is more electro-positive than iron.

- Purity of the Metal: Normally pure metal does not corrode as there is no cathode spot available to induce corrosion.

- Surface of the Metal: A rough surface corrodes readily as it collects more dirt and provides more cathode spot for corrosion. A polished surface does not corrode easily.

- Stress Corrosion: Stress in a metal surface is produced by mechanical workings such as quenching, pressing, bending and riveting, improper heat treatment etc. The portion subjected to more stress acts as anode and other portion act as cathode. This leads to the formation of stress corrosion.

  Stress corrosion is noted in fabricated articles of certain alloys like high zinc brasses and nickel brasses. Caustic embrittlement noted in boilers is a typical example for the stress corrosion which is due to the attack of alkali present in water on stressed boiler metal.

- Anode to Cathode Area Ratio: When a bigger cathode area covers a smaller anode area severe corrosion is noted in the anode spot. This is called erosion. It is frequently encountered in piping agitators, condenser tubes etc. where turbulent flow of gases and vapours remove the coated surfaces resulting in differential cells. Removal of surface coatings can also be caused by rubbing or striking activities of solids on the coated surfaces.

- Physical State of a Metal: The rate of corrosion is influenced by grain size, orientation of crystals, stress etc. The smaller the grains size of the metal greater the rate of corrosion.

2. Factors Connected with the Environment

- Nature of the atmosphere.

- Temperature of the atmosphere.

- pH of the atmosphere.

- Amount of moisture in the atmosphere.

- Amount of oxygen in the atmosphere.

- Amount of chemical fumes in the atmosphere etc.

Examples:

- Provide space for all numbers Buried pipelines and cables passing from one type of soil to another suffer corrosion due to differential aeration.

- The lead pipe lines passing through clay and then through sand.

- The lead pipe line passing through clay get corroded because it is less aerated than sand.

(A) In some cases the corroded product sticks to the surface and absorbs more moisture. This induces further corrosion.

Ex: Rusting of iron, as rust formed over iron absorbs more moisture, rate of rusting of iron increases.

(B) In some cases the corroded product acts as the protective coating which prevents further corrosion.

Ex: Aluminium oxide formed over the surface of aluminium prevents further corrosion and act as a protective coating. This is the basic principle of anodization.

(C) In some other cases the corroded product falls out of position exposing the fresh metal surface for further corrosion.

Ex: Magnesium oxide formed over the surface of the magnesium falls out of position exposing a fresh surface for further corrosion.

3.6.2 Protection from Corrosion

As the corrosion process is very harmful and losses incurred are tremendous, it becomes necessary to minimize or control corrosion of metals. Corrosion can be stopped completely only under ideal conditions. But the attainment of ideal conditions is not possible. However, it is possible only to minimize corrosion considerably.

Since the types of corrosion are so numerous and the conditions under which corrosion occurs are so different, diverse methods are used to control corrosion. As the corrosion is a reaction between the metal or alloy and the environment, any method of corrosion control must be aimed at either modifying the metal or the environment.

Corrosion Inhibitors

Corrosion inhibitor is a substance which reduces the corrosion of a metal, when it is added to the corrosive environment.

Types of Inhibitors

Inhibitors are classified into three types. They are:

- Anodic inhibitors.
- Cathodic inhibitors.
- Vapour phase inhibitors.

1. Anodic Inhibitors

Anodic inhibitors are those that prevent the corrosion reaction occurring at the anode by forming an insoluble compound with the newly produced metal ions. These precipitates are adsorbed on the anode surface forming a protective film and reducing the corrosion rate.

Example: Chromates, Nitrates, Phosphates.

Though this type of control is effective it may be dangerous. Since severe local attack can occur, if some areas are uncovered.

2. Cathodic Inhibitors

In an electrochemical corrosion, the cathodic reactions are of two types, depending upon the environment.

i. In an Acidic Solution

Example: Amines, heavy metal soaps, mercaptans.

In an acidic solutions, the cathodic reaction is evolution of hydrogen.

i.e. $2\,H + 2\,e^- \rightarrow H_2$.

i. By Slowing Down the Diffusion of $H^{(+)}$ ions to the Cathode. This can be done by organic inhibitors like amines, pyridines etc., which are adsorbed at the metal surface.

ii. By Increasing the Over Voltage of Hydrogen Evolution. This can be done adding antimony and arsenic oxides, which deposit adherent film of metallic arsenic or antimony at the cathodic areas.

ii. In a Neutral Solution

Example: Sodium Sulphite $\left(Na_2SO_3\right)$, Hydrazine (N_2H_4). In a neutral solution, the cathodic reaction is,

$$H_2O + \frac{1}{2}O_2 + 2\,e^- \rightarrow 2\,OH^-$$

The corrosion can be reduced in two ways:

(i) By eliminating the oxygen from the neutral solution, thereby formation of OH- ions are inhibited. This can be done by adding reducing agents like Na_2SO_3, N_2H_4 etc.

(ii) By eliminating the OH- ions from the neutral solution. This can be done by adding Mg, Zn or Ni salts. These react with OH- ions form insoluble hydroxides which are deposited on the cathode forming more or less impermeable self-barriers.

3. Vapour Phase Inhibitors (VPI)

Vapour phase inhibitors are organic inhibitors which readily vapourize and form a protective layer on the metal surface. VPI are used in the protection of storage containers, packing materials, etc.

Example: Benzotriazole.

3.6.3 Design and Material Selection

- Differential metal corrosion can be reduced by avoiding the use of two dissimilar metals. If there use is unavoidable, the position of the metals chosen shall be as close as possible in the galvanic series.

- While joining the metals, care should be taken not to leave gaps between them, where some liquid or air can be trapped resulting in differential aeration corrosion.

- Stress corrosion may be avoided by annealing the equipment.

- To reduce the rate of corrosion, anodic material must be as large as possible while the cathode should be small.

3.6.4 Cathodic Protection

Cathodic Protection (CP) is a technique used to control the corrosion of a metal surface by making it the cathode of an electrochemical cell. A simple method of protection connects the metal to be protected to a more easily corroded "sacrificial metal" to act as the anode. The sacrificial metal then corrodes instead of the protected metal.

For structures such as long pipelines where passive galvanic cathodic protection is not adequate an external DC electrical power source is used to provide sufficient current.

Cathodic Protection Method

Sri Humphrey Davy 's pioneering work (1824) on protecting the copper sheathing on wooden hulls in the British Navy by sacrificial zinc and iron anodes is considered to be one of the oldest example of application of cathodic protection.

Copper-sheathed ship hulls are protected by sacrificial blocks of iron. Zinc alloy is used as sacrificial anode. Galvanizing is a typical example of sacrificial anode to protect steels.

The metallic surfaces which are exposed to an electrolyte have a multitude of microscopic anodic and cathodic sites. Where anodes are more electronegative than the cathodes, a potential difference is created between them, which helps for corrosion to occur.

The function of cathodic protection is to reduce potential difference between cathodes and anodes to a neglected value. This reduction is due to the polarization of cathodes to the potential of most active anodes. In this way, corrosion current is mitigated according to Ohm's law.

The cathodic protection can be accomplished by sending a current into the structure from an external electrode and polarizing cathodic sites in an electronegative direction.

In order to achieve adequate CP, protected structure must be polarized to a certain value. The polarized potential is measured with respect to a certain reference electrode. A copper/copper sulfate reference electrode (CSE) is the most common electrode used in soil and freshwater.

There are two types of criteria for determining the cathodic protection. Either one may be used depending on the circumstances, although the first is considered superior in most of the cases.

The Potential Criterion

The polarized potential of the protected structure is to be equal to or more negative than -850 millivolts (mV) with respect to CSE.

The Polarization Shift Criterion

The protected structure is to be polarized by 100 mV with respect to CSE from its corrosion potential.

Principle

The principle involved in the cathodic protection is to force the metal to behave like a cathode.

The important cathodic protections are:

• Sacrificial anodic protection.

- Impressed current cathodic protection.

Sacrificial Anodic Protection Method

In this method, the metallic structure to be protected is made cathode by connecting it with more active metal (anodic metal). So that all the corrosion will concentrate only on the active metal. The artificially made anode thus gradually gets corroded protecting the original metallic structure.

Hence this process is otherwise known as sacrificial anodic protection. Aluminium, Zinc, Magnesium are used as sacrificial anodes.

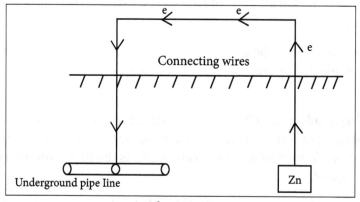

Anode protection.

Requirements of galvanic sacrificial anodes:

- Potential between the anode and corroding metal structure should be large enough to overcome the anode-cathode cells.

- Sacrificial anode to have sufficient Electrical Energy Content (EEC) which predicts its life.

- Good current efficiency relevant to anodic corrosion.

EC can be expressed and estimated as ampere hours/weight (kg or lb). Example: Pure Zinc that possesses high EEC of 372 ampere hour/pound. This means if the zinc sacrificial anode has to discharge continuously one ampere, one pound of its weight would be consumed in 372 hours. Lower current discharge will prolong its life further.

Impressed Current Cathodic Protection

In this method an impressed current is applied in the opposite direction of the corrosion current to nullify it and the corroding metal is converted from anode to cathode.

This can be done by connecting negative terminal of the battery of the metallic structure, to be protected and positive terminal of the battery is connected to an inert anode.

Inert anodes used for this purpose are graphite, platinized titanium. The anode is buried in a 'back fill'. The 'back fill provides good electrical contact to anode.

Applications of Impressed Current Protection

Structures like tanks, pipelines, transmission line towers, underground water pipe lines, ships, etc. can be protected by this method.

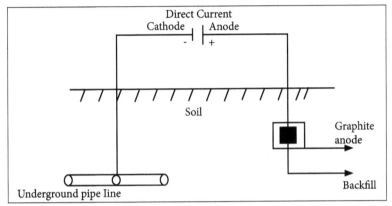

Impressed current protection.

Application of Cathodic Protection Systems on Buried Pipelines

Pipelines are used for the purpose of transporting water, petroleum products, natural gas and other utilities. There's a huge network of piping systems which are in use in every country all over the world. Pipelines may be onshore or offshore and are prone to corrosion in both cases. If corrosion isn't mitigated it can lead to dangerous and expensive damage.

There are several corrosion control techniques which are used on pipelines, cathodic protection is one of them. It can be applied either to the coated pipelines to mitigate the corrosion attack on areas where coating quality may be poor. It is also used on bare pipelines.

Both types of CP can be applied to the buried pipelines. The application of either of these types depends on several factors, such as the required current, soil resistivity and the area to be protected.

Cathodic protection aims to polarize a pipeline to a minimum potential of -850 mv, for carbon steel and for sufficing cathodic protection. The polarized potential is to be measured through test stations, which has to be installed at the following locations along the route of pipeline:

- At points of electrical isolation.

- At crossings with foreign structures.

- At frequent intervals (e.g. < 2 km / 1.24 miles).

- At casings.

- At some galvanic anode locations.

- At the location of stray current discharge to earth.

- Near sources of electrical interference.

Problems Created by Cathodic Protection

In large pipeline networks, there are a lot of crossings, parallelism and approaches, where in the pipeline has its applied cathodic protection system. DC interference may occur between pipelines, which results in accelerating corrosion.

In order to overcome this problem, pipelines can be electrically coupled, either directly or through a resistance. For coated pipelines, where the applied coating quality is poor, cathodic disbondment may occur due to high cathodic protection levels.

Higher temperatures can also promote cathodic disbondment. High pH environments are also a concern in terms of stress-corrosion cracking. In such cases, the polarized potential of the pipeline must be kept at a minimum value of -850 mV.

Remember, cathodic protection is just one of the several methods that are used to prevent corrosion, not just in pipelines, but in ships, offshore oil platforms and other steel structures. Whether it's the best application for the job or the only one to be used, is often specific to the structure being protected.

3.7   Protective Coatings: Surface Preparation

Protective Coatings

Corrosion of metal surfaces is a common phenomenon. To protect a metal surface from the corrosion, the contact between the metal and the corrosive environment is to be cut off. This is done by coating the surface of the metal with a continues, non-porous material, inserted to the corrosive atmosphere. Such a coating is referred to as surface coating or protective coating.

In addition to protective action, such coatings also give a decorative effect and reduce wear and tear.

Objectives of Coating Surfaces

- To prevent corrosion.

- To enhance the wear and scratch resistance.

- To increase hardness.

- To insulate electrically.

- To insulate thermally.

- To impart decorative colour.

Surfacing coatings made up of metals are known as metallic coatings. These coatings separate base metal from the corrosive environment and also function as an effective barrier for the protection of base metals.

Surface coating is a mixture of film-forming materials plus pigments, solvents and other additives, which, when applied to a surface and cured or dried, yields a thin film that is functional and often decorative.

Surface coatings include the paints, drying oils and varnishes, synthetic clear coatings and other products whose primary function is to protect the surface of an object from the environment.

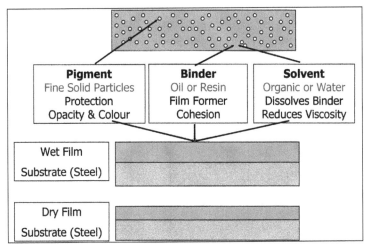

Paints-constituents.

Surface Preparation

Surface preparation is the essential first stage treatment of a substrate before the application of any coating. The performance of a coating is significantly influenced by its ability to adhere properly to the substrate material. It is generally well established that correct surface preparation is the most important factor affecting the total success of surface treatment.

The presence of even small amounts of surface contaminants, oil, grease, oxides etc. can physically impair and reduce coating adhesion to the substrate. The chemical contaminants that are not readily visible, such as chlorides and sulphates, attract mois-ture through coating systems resulting in premature failure. The importance of a chem-ically clean substrate to provide the best possible contact surface for the applied coating cannot be over-emphasized.

The adhesion of coating materials such as zinc by hot dip galvanizing is particularly good due to the metallurgical alloying with the steel surface. In this process, the molten zinc reacts with the steel to form a series of iron / zinc alloy layers of varying composi-tion to form an intimate bond with the substrate.

Alternatively, metal coatings applied by thermal spraying require a coarse surface pro-file to maximize the coating bond which is principally by mechanical keying. Organic paint coatings adhere to the surface mainly by the polar adhesion which is aided by mechanical adhesion which is especially important for thick coating films. For all types of coatings, the surface condition of the substrate is critical in terms of coating perfor-mance and durability.

Organic Coatings

Paint is a mechanical dispersion of one or more fine pigments in a medium (thinner + vehicle). When paint is applied to metal surface the thinner evaporates. The vehicle undergoes slow oxidation to form a pigmented film.

Properties of Paint

A good paint should have the following properties:

- It should have high hiding power and the required colour.

- It should be able to resist the atmospheric conditions to which it will be put.

- The films that are produced should be washable.

- It should resist corrosion.

- It should have the necessary consistency (property to resist permanent change of shape) for a particular purpose for which the paint is to be used.

- The film produced by the application of paint on a surface should have gloss.

Purpose of Paint

The purpose of using a paint is as listed below:

- To avoid loss of metal due to corrosion.

Paint protects the metal surfaces from the corrosive effects of weather (sun, wind, rain, frost, atmospheric pollution & other natural elements), heat, moisture, gases etc.

- Delays in rusting.
- It also provides:
  - aesthetic look to materials.
  - a smooth surface for easy cleaning.

The paints can be classified into two types:

- Classification of paints based on physical type.
- Classification of painting products based on their functions.

Classification Based on Physical Type

- Solvent-borne paints.
- Water- borne paints.
- High solid paints.
- Powder coatings.
- Radiation curable coatings.

Classification Based on Function

- Paint
- Varnish
- Enamel
- Primer

Constituents of Paint

Pigment, Vehicle, Thinner, Drier, Filter, Plasticizer, Anti skinning agent.

1. Pigment: It is a solid that gives colour to the paint.

Example: White pigment - White lead, TiO_2.

Functions:

- To give colour and opacity to the film.
- To provide strength to the film.

- To protect film by reflecting UV rays.

- To provide resistance to abrasion and weather.

2. Vehicle Oil: It is the film-forming liquid. It holds the ingredients of the paint. It is a non-volatile high molecular weight fatty acid of vegetable or animal.

Example: Lin seed oil, Castor oil.

Functions:

- To hold the pigment on the surface.

- To form a protective layer by oxidation and polymerization.

- To impart water repellence, toughness and durability of film.

- To improve adhesion of film.

3. Thinner: It is added to reduce the viscosity of the paint for easy application on the surface. It easily evaporates after paint is applied.

Example: Turpentine.

Functions:

- To reduce viscosity of paint.

- To dissolve vehicle and other additives.

- To suspend the pigments.

- To increase elasticity of film.

- To increase penetration of vehicle.

- To improve drying of film.

4. Drier: It is a substance used to speed up drying of the paint.

Example: Metallic soap, linoleate of CO.

Functions:

- To act as oxygen carrier or catalyst.

- To provide oxygen essential for oxidation and polymerization of drying oil.

5. Extender or Filter: These are pigments that form bulk of the paint.

Example: Gypsum.

Functions:

- To reduce cost of paint.
- To prevent shrinkage and cracking of film.

6. Plasticizer: It is added to the paint to provide elasticity to the film and prevent its cracking.

Example: Triphenyl phosphate, Tricresyl phosphate.

7. Anti skinning Agent: It is a chemical added to the paint to prevent gelling and peeling of the point.

Example: Polyhydroxy Phenols.

3.7.1 Metallic (Cathodic and Anodic) Coatings

Metal coating: Corrosion of metal can be controlled by isolating them from the corrosive atmosphere. This can be done by covering the metal (base metal) with a layer of another metal. This process is known as metal coating.

1. Anodic Metal Coating (Sacrificial Coating)

It is produced by coating a base metal with more active metal or more anodic to the base metal for e.g: Iron is coated with Zn, Mg, Al etc. One of the important properties of this type of coating is that, even if the coating is ruptured, the base metal does not undergo corrosion.

The exposed part of the base metal is cathodic with respect to the coating metal and coating metal only undergoes corrosion there by protecting the base metal. The protection is there as long as coating is there. Galvanization is one of the best example in anodic coating.

Galvanization

It is a process of coating a base metal (iron) with zinc (Zn) metal. This process is usually carried out by hot dipping method.

- The base metal surface is washed properly with organic solvents to remove any organic matter (like oil, grease etc) on the surface.
- It washed with dil.H_2SO_4 to remove any inorganic matter (like rust).
- Finally the base metal is well washed with water and air-dried.
- The base metal then dipped in a bath of molten zinc maintained at 425-430°C and covered with a flux of NH_4Cl to prevent the oxidation of molten zinc. Then

excess zinc on the surface is removed by passing through a pair of hot rollers so that a proper thin coating is obtained.

Application

Galvanized articles are mainly used in roofing sheets, fencing wire, buckets, bolts nuts, pipes and tubes etc. but galvanized articles are not used for preparing and storing food stuffs. Since zinc dissolves in dilute acids and become toxic.

2. Cathodic Metal Coating

These are the coating produced by coating a base metal with more cathodic (noble metal) for e.g.: iron is coated with tin, nickel and Cu. But these coatings provides protection only when it is undamaged and absolutely free from gaps otherwise rapid corrosion of the base metal takes place as a result of the formation of large cathodic and small anodic area.

Tinning is the best example to explain cathodic coating.

Tinning

It is a process of coating base metal (iron) with tin (Sn). It can be carried out by hot dipping method.

- The iron sheet is (base metal) first washed thoroughly with organic solvents to remove any organic matters.

- Then treated with dil.H_2SO_4, to remove rust.

- Finally it is washed well with water and air-dried.

- It is then passed through $ZnCl_2$ flux, so that molten tin can adheres properly on the metal surface

- Then base metal is passed through tank that contains molten tin.

- Finally passed through a series of rollers immersed in palm oil. So that uniform, undamaged, continuous deposit of tin takes place. Tinning will provide complete protection against corrosion if it covers the surface completely.

Application

Tinned articles are largely used in the manufacturing of containers used for storing foodstuffs, copper utensils are coated with tin so that contamination of food with copper can be prevented.

3.7.2 Methods of Application on Metals (Galvanizing, Tinning, Electroplating, Electroless Plating)

The metal which is coated upon is known as base metal. The metal applied as coating is referred to as coat metal.

The different methods used for metal coating are:

- Hot dipping.
  - Galvanization
  - Tinning
- Metal spraying.
- Cladding.
- Cementation
  - Sherardizing - Cementation with Zinc powder is called Sherardizing.
  - Chromizing - Cementation with 55% Chromium powder and 45% Alumina is called chromizing.
  - Calorizing – Cementation with Aluminium and Alumina powder is called Calorizing.
- Electroplating or electrodeposition.
- Electroless plating.

1. Hot Dipping

In the process of hot dipping, the metal to be coated is dipped in the molten bath of the coating metal. Such hot dip coatings are generally non-uniform. The common examples of hot dip coatings are galvanizing and tinning.

i. Galvanization

The process of coating a layer of zinc on iron is called galvanizing. The iron article is first pickled with dilute sulphuric acid to remove traces of rust, dust, etc. at 60-90°C for about 15 -20 minutes. Then this metal is dipped in a molten zinc bath maintained at 430°C

The surface of the bath is covered with ammonium chloride flux to prevent oxide formation on the surface of the molten zinc. The coated base metal is then passed through rollers to correct the thickness of the film.It is used to protect roofing sheets, wires, pipes, tanks, nails, screws, etc.

ii. Tinning

The coating of tin on iron is called tin plating or tinning. In tinning, base metal is first pickled with dilute sulphuric acid to remove surface impurities. Then it is passed through molten tin covered with zinc chloride flux.

The tin coated article is passed through a series of rollers immersed in a palm oil bath to remove the excess tin. Tin-coated utensils are used for storing food stuffs, pickles, oils, etc. Galvanizing is preferred to tinning because tin is cathodic to iron, whereas zinc is anodic to iron. So, if the protective layer of the tin coating has any cracks, iron will corrode.

If the protective layer of the zinc coating has any cracks, iron being cathodic does not get corroded. The corrosion products fill up the cracks, thus preventing corrosion.

Differences between Galvanizing and Tinning

Galvanizing	Tinning
A process of covering iron with a thin coat of 'Zinc' to prevent it from rusting.	A process of covering iron with a thin coat of 'tin' to prevent it from corrosion.
Zinc protects the iron sacrificially. (Zinc undergoes corrosion).	Tin protects the base metal without undergoing any corrosion (non sacrificially).
Zinc continuously protects the base metal even if broken at some places.	A break in coating causes rapid corrosion of base metal.
Galvanized containers cannot be used for strong acidic food stuffs as Zinc becomes toxic in acidic medium.	Tin is non-toxic in nature of any medium.

2. Cladding

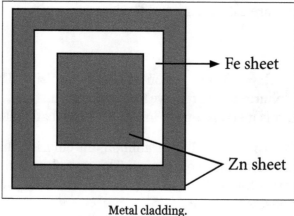

Metal cladding.

Cladding is applicable to sheet metals where in a metal sheet which has to be prevented from corrosion is cladded between two anodic metal sheets and heated to high temperature and rolled over at high pressure to form one piece which is corrosion resistant.

For example, iron sheet is sandwiched between two zinc sheets shown in figure. Likewise aluminium sheets can be cladded by Duralumin, an alloy of magnesium, aluminium and copper, which is used in aircraft industry to prevent corrosion.

This method finds application in aircraft industry where they use different types of metals and alloys.

3. Metal Finishing

The materials such as metals/alloys are required for various engineering applications. These materials should be ideal and must meet several requirements like resistance to corrosion, wear resistance, mechanical properties, etc.

It is impossible to have all these properties in a single metal. Hence, to improve the lacking properties in these materials, metal finishing is one of the method employed for the purpose.

Metal finishing is the process carried out to modify the surface properties of a metal by electro deposition of a layer of another metal on substrate.

Technological importance of metal finishing are:

- To increase the corrosion resistance.

- To increase thermal resistance.

- To increase the durability of the metal.

- To impart hardness.

- To provide electrical and thermal conducting surface.

- Manufacturing electrical and electronic components such as PCB, capacitors etc.

- Decorative purpose.

Important techniques of metal finishing are:

- Electroplating.

• Electroless plating.

4. Electroplating

Gold Electroplating

It is the process of applying a thin layer of gold onto a desired metal material, normally that of copper or silver, through the process of electroplating.

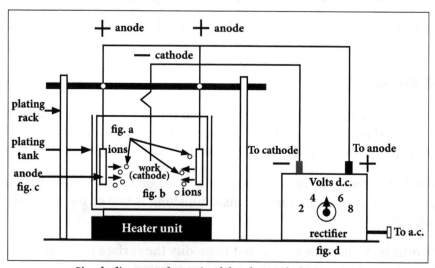

Simple diagram schematic of the electroplating process.

The Benefits/features of Gold Electroplating

Gold electroplating, when used in jewelry manufacturing, allows the manufacturer to provide the market with jewelry that appears to be made of pure gold at a cheaper price as the jewelry is only plated with gold.

This process can also provide benefits to the electronics industry when used. It allows the electronic device to be more conductive and more resistant to wear, thus performing better and lasting longer.

Typical Application of Gold Electroplating

It is commonly used in manufacturing jewelry by plating silver and in electronics in electrical connectors and printed circuit boards which forms an electrically conductive layer on copper that is also resistant to corrosion.

The Process of Gold Electroplating

When gold electroplating is used on silver in the manufacturing of jewelry, a copper and nickel layer must be deposited onto silver before the gold electroplating occurs (pre-treatment). The reason for this is that if no intermediate layer were provided the silver atoms would, over time, diffuse through the gold plating and cause what is known as tarnishing.

The copper and nickel layers slow down this process. With electronics, an intermediate layer of the nickel is used as copper will, like silver, diffuse through the gold layer and cause tarnishing and a sulfide/oxide layer to form.

Electroplating of Nickel (Watt's Bath) or Watts Nickel Plating Solutions

It is a process of depositing the nickel on a metal part. Parts to be plated must be clean and free of dirt, corrosion and defects before plating can begin. To clean and protect the part during the plating process a combination of heat treating, cleaning, masking, pickling and etching may be used.

Once the piece has been prepared it is immersed into an electrolyte solution and is used as cathode. The nickel anode is dissolved into the electrolyte in form of nickel ions.

Ions travel through the solution and deposit on the cathode Watts solution was developed by Oliver P. Watts in 1916. Plating operation in Watts solutions is low cost and simple.

Bath Composition	
Nickel Sulphate	$NiSO_4 6H_2O$: 32-40 oz/gal (240-300 g/l)
Nickel Chloride	$NiCl_2 6H_2O$: 4-12 oz/gal (30-90 g/l)
Boric Acid	H_3BO_3: 4-6 oz/gal (30-45 g/l)
Operating conditions	
Temperature	105-150°F (40-65°C)
Cathode current density	20-100 A/ft² (2-10 A/dm²) pH: 3.0-4.5
Mechanical Properties	
Tensile strength	50000-70000 psi (345-485 MPa)
Elongation	10-30%
Hardness	130-200 HV
Internal stress	18000-27000 psi (125-185 MPa)

Electroplating of chromium: It is carried out on the pre treated Ni or Cu articles.

Chromium plating	Decorative chromium
Anode	Insoluble anodes like P_b-S_b or P_b-S_n coated with PbO_2 etc
Cathode	Object to be plated
Bath composition	chromic acid and H_2SO_4 in the ratio 100:1

Current density(mA/cm²)	100-200
Temperature(°C)	45-55
pH	2-4
Current efficiency (%)	8-12
Applications	Provides durable and good decorative finish on automobiles, surgical instruments

$$Cr_2O_7^{2-} + 14H^+ + 6e^- \rightarrow 2Cr^{3+} + 7H_2O$$

$$Cr_3^+ + 3e \rightarrow Cr$$

In chromium plating sulphate ion provided by the sulphuric acid act as a catalyst. Cr is present in the hexavalent state Cr(VI) as CrO_3 in the bath solution. This is converted into trivalent Cr(III) by a complex anodic reaction in the presence of sulphate ions and then coated, as Cr on the cathode surface the amount of Cr(III) ions should be restricted in order to get proper deposit.

Insoluble anodes are used to maintain the Cr(III) concentration as they oxidize Cr(III) to Cr(VI). The chromium anodes are not used in Cr plating because Cr metal passivates strongly in acid sulphate medium and Cr anode gives Cr(III) ions on dissolution. In the presence of large concentration of Cr(III) ions, a black Cr deposit is obtained.

5. Electroless Plating

Pretreatment of Surface

1. The surface to be plated is decreased by treating with organic solvent and etched in acid. It is then activated by dipping it in $SnCl_2$ in HCL followed by $PdCl_2$ in HCL.

2. Plating bath composition,

Solution of Copper sulphate $(CuSO_4 \cdot 5H_2O) - 12$ g/L.

Reducing agent: Formaldehyde - 8 g/L.

Buffer: Sodium hydroxide - 15g/L + Roschells salt-14g/L.

Complexing agent: EDTA-15g/dm³.

Temperature: 25 °C.

pH – 11.

Reactions

Atanode : $2HCHO + 4OH^- \rightarrow 2HCOO- + 2H_2O + H_2 + 2e^-$

At cathode: $Cu^{2+} + 2e^- \rightarrow Cu$

Net reaction: $2HCHO + 4OH^- \rightarrow 2HCOO^- + 2H_2O + H_2 + Cu$

Applications

- For decorative appearance.

- In preparation of PCB.

Preparation of PCB by Electroless Plating

- Base material is made up of glass reinforced plastic or a epoxy polymer.

- Base material which is double sided is covered by thin layers of electroformed copper.

- Selected areas are protected by photo resist or etch resist.

- The rest of copper is removed by etching with suitable etchant to get circuit pattern.

- The contact between two sides is done by drilling holes at required points, followed by plating of copper in the holes by the electroless plating.

Electroless Plating of Nickel

Principle

Electroless plating is a technique of depositing a noble metal on a catalytically active surface of the metal to be protected by using a suitable reducing agent without using electrical energy.

The reducing agent reduces the metallic ions to metal which gets plated over the catalytically activated surface giving a uniform thin coating.

$$\text{Metal ions} + \text{reducing agent} \rightarrow \text{Metal} + \text{oxidized products.}$$

Step 1

Pretreatment and activation of the surface:

The surface to be plated is first degreased by using organic solvents followed by acid treatment. Activation depends on the type of the metal (or) alloy (or) non-metal used.

Example:

- The surface of the stainless steel is activated by dipping in hot solution of 50% dil.H_2SO_4.

- The surface of Mg alloy is activated by their coating of Zn and Cu over it.

- Metals and alloys like Al, Cu, Fe, etc. can be directed to Ni-plated without activation.

Step 2

Space Preparation of plating bath.

The plating bath consists of the following ingredients.

Nature of the compound	Name of the compound	Quantity (g/l)	Function
1. Coating solution	$NiCl_2$	20	Coating metal
2. Reducing agent	Sodium hyophosphite	20	Metalions
3. Complexing agent cumexhalant	Sodium succinate	15	Improves quality
4. Buffer	Sodium acetate	10	Control pH
5.Optimum pH	4.5	-	-
6.Optimum temperature	93°C	-	-

Step 3

Procedure for plating.

The preheated object is immersed in the plating bath for required time. During which the following reduction reaction will occur and the nickel gets coated over the object.

Reactions

At cathode : $Ni^{2+} + 2e^- \rightarrow Ni$

At anode : $H_2PO_2 + H_2O \rightarrow H_2PO_3^- + 2 H^+ + 2e^-$

Net reaction : $Ni^{2+} + H_2PO_2 - + H_2O \rightarrow Ni + H2PO_3^- + 2H^+$.

Applications

- Electroless Ni-plating is extensively used in electronic appliances.

- Electrodes Ni-plating is used in domestic as well as automobile fields.

Differences between Electro Plating and Electroless Plating

Electroplating	Electroless plating
1. Electrical energy is required to bring about plating.	Electrical energy is not required to bring about plating.
2. Oxidation takes place at anode & reduction at cathode.	Both oxidation & reduction takes place on catalytically activated substrate.

3. Object to be plated acts as cathode.	Object to be plated which is catalytically active acts as cathode.
4. Metal is the anodic reactant.	Reducing agent is anodic reactant.
5. Anodic reaction is $M \rightarrow Mn^+ + ne^-$	Anodic reaction is, Reducing agent $\rightarrow$ oxidized product $+ ne^-$
6. Not economical.	Most economical.
7. Throwing power is low.	Throwing power is high, Lower case So deposit is more uniform over a article, irrespective of its shape and size.
8. Done only over metallic surfaces.	Done over non metallic surfaces like, plastics and other polymers.
9. Pure metal or alloy is deposited.	Deposit may be contaminated with oxidized product or reducing agent.

Chemistry of Advanced Materials

4

4.1 Nano Materials: Sol-gel Method and Chemical Reduction Method of Preparation and its Characterization

Nano materials are defined as the materials with at least one external dimension in the size range from approximately 1-100 nanometers. Nano particles are the objects with all three external dimensions at the nanoscale.

Nano particles that are naturally occurring (e.g., volcanic ash, soot from forest fires) or are the incidental byproducts of combustion processes (e.g., welding, diesel engines) are usually physically and chemically heterogeneous and often termed ultrafine particles.

Salient Properties of Nano Materials

1. Melting Points

Nano materials have a significantly lower melting point and appreciable reduced lattice constants. This is due to huge fraction of surface atoms in the total amount of atoms.

2. Optical Properties

Reduction of material dimensions has pronounced effects on the optical properties. Optical properties of nano materials are different from bulk forms. The change in optical properties is caused by two factors.

- The quantum confinement of electrons within the nano particles increases the energy level spacing.

- Surface plasma resonance, which is due to smaller size of nano particles than the wavelength of incident radiation.

3. Magnetic Properties

Magnetic properties of nano materials are different from that of bulk materials. Ferromagnetic behaviour of bulk materials disappear when the particle size is reduced and transfer to super para-magnetics. This is due to the huge surface area.

4. Mechanical Properties

The nanomaterials have less defects compared to bulk materials, which increases the mechanical strength.

- Mechanical properties of polymeric materials can be increased by the addition of nano-fillers.

- As nano materials are stronger, harder and more wear resistant and corrosion resistant, they are used in spark plugs.

Applications of Nanomaterials

Nanomaterials posses unique and beneficial, physical, chemical and mechanical properties, they can be used for a wide verity of applications.

Material Technology

- Nanocrystalline aerogel are light weight and porous, so they are used for insulation in offices homes, etc.

- Cutting tools made of nanocrystalline materials are much harder, much more wear- resistance and last stronger.

- Nanocrystalline material sensors are used for smoke detectors, ice detectors on air craft wings, etc.

- Nanocrystalline materials are used for high energy density storage batteries. Nanosized titanium dioxide and zinc dioxide are used in sunscreens to absorb and reflect ultraviolet rays.

- Nan coating of highly activated titanium dioxide acts as water repellent and antibacterial. The hardness of metals can be predominately enhanced by using nanoparticles.

- Nanoparticles in paints change colour in response to change in temperature or chemical environment and reduce the infrared absorption and heat loss.

- Nanocrystalline ceramics are used in automotive industry as high strength springs, ball bearings and valve lifters.

Information Technology

- Nanoscale fabricated magnetic materials are used in data storage Nano computer chips reduce the size of the computer.

- Nanocrystalline starting light emitting phosphors are used for flat panel displays.

- Nanoparticles are used for information storage.

- Nanophotonic crystals are used in chemical optical computers.

Biomedicals

- Bio sensitive nano materials are used for ragging of DNA and DNA chips.

- In the medical field, nano materials are used for disease diagnosis, drug delivery and molecular imaging.

- Nanocrystalline silicon carbide is used for artificial heart valves due to its low weight and high strength.

Energy Storage

- Nanoparticles are used hydrogen storage.

- Nanoparticles are used in magnetic refrigeration.

- Metal nanoparticles are useful in fabrication of ionic batteries.

Various Type of Synthesis Involved in the Preparation of Nanomaterials

Nanomaterials are synthesized in two methods:

- Top-down methods.

- Bottom-up methods.

Top-down Methods Bottom-up Methods

- Laser ablation.

- Chemical vapour deposition.

- Electrodeposition.

1. Laser Ablation Method

In laser ablation, high-power laser pulse is used to evaporate the matter from the target. The stoichiometry of the materials is preserved in the interaction.

2. Chemical Vapor Deposition

It is a process of chemically reacting a volatile compound of a material with other gases to produce a non-volatile solid that deposits automatically on a suitably placed substrate. CVD requires activation energy to proceed. This energy can be provided by several methods.

3. Electrodeposition

Template assisted electrodeposition is an important technique for synthesized metallic nanomaterials with controlled shape and size. Arrays of nano-structured materials with specific arrangements can be prepared by this method, using as active template as a cathode in an electro chemical cell.

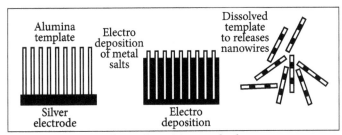

Electrodeposition method.

The electrodeposition method consists of an electrochemical cell. The cell usually contains a reference electrode, a specially designed cathodes and an anode. The cathode substrate on which electro-deposition of the nano-structures takes place can be made of either non-metallic or metallic materials. By using the surface for the cathode as template various desired nano-structures.

Synthesis of nanomaterial precipitation and thermolysis methods. [Bottom–up (or) chemical (or) soft method (or) small to big method].

This method is carried out by the following:

Process

- Precipitation
- Thermolysis

1. Precipitation (PPL)

Nano particle generally are synthesized by the precipitation reactions between the reactant in presence of water soluble inorganic stabilizing agent.

Example:

i. Precipitation of BaSO$_4$ Nano-particles

10 gm of sodium hexameta-phosphate (stabilizing agent was dissolved in 80ml of distilled water in 250 ml beaker with constant stirring. Then 10ml of $Ba(NO_3)_2$

solution. The resulting solution was stirred for one hour. Precipitation occurs slowly. The resulting precipitation was then centrifuge washed with distilled water and vaccum dried.

$$Ba(NO_3)_2 + Na_2SO_4 \rightarrow BaSO_4 \downarrow + 2NaNO_3$$

Bulk $BaSO_4$ is obtained in the absence of stabilizing agent.

ii. Precipitation by Reduction

Reduction of metal salt to the corresponding metal atoms. These atoms act as nucleation centres leading to formation of atomic clusters. These clusters are surrounded by stabilizing molecule that prevent the atoms agglomerating.

2. Thermolysis

Thermolysis is characterized by subjecting the metal precursors at high temperature together with a stabilizing agent. Nano-particles show an increase in size relating to the temperature rise. This is due to the elimination of stabilizing molecule generating a greater aggregation of the particle. The following equation gives the example for the process of thermolysis.

i. Hydrothermal Synthesis

It involves crystallization of substance from high temperature aqueous solution at high vapour pressure, hydrothermal synthesis is generally performed below the supercritical temperature of water (374°C).

Method

It is performed in an apparatus consisting of a steel pressure vessel caused autoclave in which a nutrient is supplied along with water a gradient of temperature is maintained.

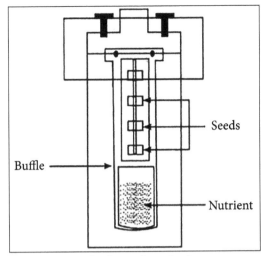

Hydrothermal synthesis.

At the opposite ends of growth chamber, so that the hotter end dissolves the nutrient and the cooler end causes seeds to take additional growth.

ii. Solvo Thermal Synthesis

It involves the use of solvent under high temperature (between 1000°C) and moderate to high pressure (1 atm to 10000 atm) that facilitate the interaction of precursors during synthesis.

Method

A solvents is mixed with certain metal precursors and the solution mixture is placed.

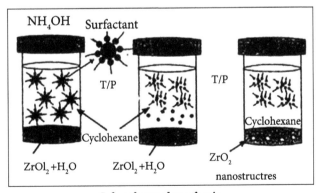

Solvo thermal synthesis.

In an autoclave kept at relatively high temperature and pressure in an oven to carry out the crystal growth. The pressure generated in vessel due to solvent vapour elevates the boiling point of the solvent.

Example for solvent: Ethanol, Toluene, Cyclohexane etc.

Electrodeposition

Template assisted electrodeposition is an important technique for synthesizing metallic nano-materials with controlled shape and size.

The arrays of nano-structured materials with specific arrangement can be prepared by this method. Using an active template as a cathode in an electrochemical cell.

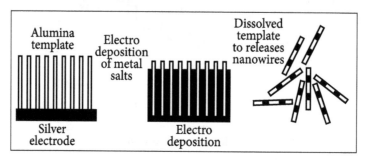

The electrodeposition method consists of an electrochemical cell. The cell generally contains a reference electrode, a specially designed cathodes and an anode. The cathode substrate on which electro deposition of nano-structure takes place can be made of either non-metallic or metallic materials.

By using the surface of the structures can be synthesized for specific applications. Chemical vapour deposition is a well-known process in which a solid is deposited on a cooled surface via chemical reaction from vapor or gas phase.

The basic steps involved in this process are:

- Generation of vapour by boiling or subliming a source material.

- Transformation of vapour from the source to the substrate.

- Condensation of vapour on the cool substrate.

In this method, the atoms in gaseous state allowed to react homogeneously or heterogeneously depending on the applications. This method is an excellent method which is used to control the particle size, shape and chemical compositions.

This method is used to produce nano powders of oxides and carbides of metals. Production of pure metal powders is also possible using this method.

Sol-gel Method

The sol-gel process is a wet technique i.e., chemical solution deposition technique used for the production of high purity and homogeneous nano materials. In solutions the molecules of nanometer size are dispersed and move around randomly and hence the solution are clear.

In colloids the molecules of size ranging from 20 to 100 are suspended in a solvent. When mixed with a liquid is known as sol. A suspension that keeps its shape is called as gel. Thus the colloids are suspensions of colloids in liquids that keep their shape. The formation of sol-gels involves hydrolysis, condensation growth of particles and formation of networks.

Plasma Arching Method

Plasma is an ionized gas. To produce plasma, high potential difference is applied across the electrodes. An arc passes from one electrode to another electrode. The gas yields up its electrons and positive ions at anode. Positively charged ions pass to cathode pick up electrons and are deposited to form nano particles.

Using this plasma arching method, quite thin films of the order of atomic dimensions can be deposited on the surface of an electrode. This deposition is carried in vacuum or in an inert gas. By using carbon electrodes, carbon nanotubes can be formed on the surface of the cathode.

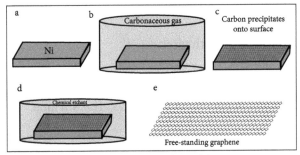

Chemical vapour deposition.

Laser Ablation

In laser ablation, high power laser pulse is used to evaporate the matter from the target. The stoichiometry of material is preserved in the interaction. The total mass ablated from target per laser pulse is referred to as the ablation rate.

CVD and Laser Ablation Techniques for the Synthesis of Nano-particles

Chemical Vapour Deposition

It is process of chemically reacting a volatile compound of a material with other gases to produce a non – volatile solid which deposits automatically on a suitable placed substrate.

CVD Reaction requires activation energy to proceed. This energy can be provided by several methods.

- Thermal CVD: In thermal CVD, the reaction is activated by high temperature above 900°C. Typical apparatus comprises of gas supply system, deposition chamber and an exhaust system.

- Laser CVD: In Laser CVD, Pyrolysis occurs when laser thermal energy of laser heats falls on an absorbing substrate.

- Plasma CVD: In Plasma CVD, the reaction is activated by plasma at temperature between 300 – 700°C.

- Photo – laser CVD: In Photo – laser CVD, the chemical reaction is induced by ultra violet radiation which has sufficient photon energy to breath the chemical bond in the reactant molecules.

Synthesis Step

- Transport of gaseous reactant to the surface.

- Adsorption of gaseous reactant on the surface.

- Catalyzed reaction occurs on the surface.

- Product diffuse to the growth sites.

- Nucleation and growth occurs on the growth site.

- Desorption of reaction products away from the surface.

CVD Reactor

The CVD reactors are of generally two types:

- Hot wall CVD.

- Cold wall CVD.

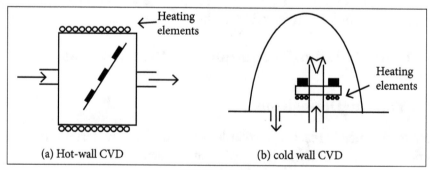

(a) Hot-wall CVD   (b) cold wall CVD

CVD Reactors.

- Hot – wall CVD reactors are usually tubular in form and heating is accomplished by surrounding the reactor with resistance elements.

- But in cold – wall CVD reactors, substrates are directly heated inductively by graphite susceptor while chamber walls are air or water cooled.

In laser ablation, high power laser pulse is used to evaporate the matter from the target. The stoichiometry of the material is preserved in the interaction. The total mass ablated from the target per laser pulse is referred to as the ablation rate.

Reaction Setup

A typical laser ablation setup in shown in figure:

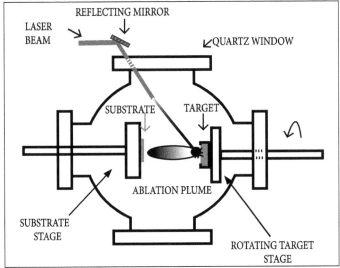

Laser ablation.

When a beam of laser is allowed to irradiate the target, a supersonic jet of particles is evaporated from the target surface. Simultaneously an inert gas such as argon, helium is allowed into the reactor to sweep the evaporated particles from the furnace zone to the colder collector.

Ablated species condense on the substrate placed opposite to the target. The ablation process takes place in a vaccum chamber either in vaccum or in the presence of some background gas.

Chemical Reduction Method of Preparation

It is interesting to note that, out of all the methods used for producing nanomaterials, chemical reduction method is increasingly attractive and offers good potential in terms of accomplishing physicochemical properties, better handling, equipment configuration and the associated cost factors involved in other processes.

The most interesting part of chemical method is that, by utilizing suitable precursors (raw materials), reducing agents and stabilizing agents (generally chemicals), the nanomaterials of relatively smaller size/narrow size distribution and desired shape can be effectively synthesized.

Moreover, the stability of the nanomaterials being dispersed into the base material (PCM) can be established very well without experiencing agglomeration (or settlement) of the particles. Chemical reduction is the most frequently applied method for the preparation of silver nanoparticles as stable, colloidal dispersions in water or organic solvents.

Generally used reducing agents are borohydride, citrate and elemental hydrogen. The reduction of silver ions $\left(Ag^+\right)$ in aqueous solution commonly yields colloidal silver with particle diameters of several nanometers. Initially, the reduction of various complexes with Ag+ ions leads to the formation of silver atoms, which is followed by agglomeration into oligomeric clusters.

These clusters eventually lead to the formation of colloidal silver particles. When colloidal particles are much smaller than the wavelength of visible light, the solutions have a yellow color with an intense band in the 380–400 nm range and other less intense or smaller bands at longer wavelength in absorption spectrum.

This band is attributed to collective excitation of the electron gas in the particles, with a periodic change in electron density at the surface (surface plasmon absorption). Controlled synthesis of silver nanoparticles is based on a two-step reduction process.

In this technique a strong reducing agent is used to produce small Silver particles, which are enlarged in a secondary step by further reduction with a weaker reducing agent. Chemical reduction of metal salts using various reducing agents in presence of stabilizer is currently of interest for preparation of metal nanoparticles.

The reducing agents such as sodium Borohydride $\left(NaBH_4\right)$, hydrazine $\left(N_2H_4\right)$, formaldehyde, etc. can be used to reduce a silver containing salt to produce the nano-silver particles.

Characterization by BET Method

The Brunauer, Emmett and Teller (BET) technique is the most common method for determining the surface area of powders and porous materials. Nitrogen gas is normally employed as the probe molecule and is exposed to a solid under investigation at liquid nitrogen conditions (i.e. 77 K).

The surface area of the solid is evaluated from the measured monolayer capacity and knowledge of the cross-sectional area of the molecule being used as a probe. For the case of nitrogen, the cross-sectional area is taken as 16.2 Å^2 / molecules.

For the case of spherical, non-porous particles, the BET surface area is related to the particle diameter (D) or radius (R) and skeletal density through,

$$\text{BET}\left[\frac{m^2}{g}\right] = \frac{4\mathring{o}R^2}{\frac{4}{3}\mathring{o}R^3\tilde{n}_s} = \frac{3}{\tilde{n}_sR} = \frac{6}{\tilde{n}_sD}$$

Nanoparticles are being incorporated into a wide range of products including drugs, personal care products, plant growth regulators, specialty additives, medical devices and nutritional products.

The high surface area systems can be attained either by fabricating small particles or clusters where the surface-to-volume ratio of each particle is high or by creating materials where the void surface area (pores) is high compared to amount of bulk support material.

Small particles such as nanoparticles have a high surface-to-volume ratio potentially resulting in increased surface reactivity, increased rate of dissolution, altered bioavailability and most importantly, a changed toxicity profile. Linsinger et al. of the Joint Research Council published a report in 2012 entitled 'Requirements on Measurements for the Implementation of the EC Definition of the Term Nanomaterial'.

In this comprehensive report Linsinger et al conclude that according to the definition set forth by the EU, specific surface area by BET surface area measurements can be used to positively classify a material as a nanomaterial. BET surface area measurements may also be highly relevant when considering the toxicological aspects of novel materials.

Characterization by Transmission Electron Microscope (TEM) Method

TEM is a very powerful tool for material analysis, which help in revealing the size and morphology of the particles, crystal structure and determination of lattice spacing (selected area diffraction pattern), defects, etc.

TEM analysis reside a high energy beam of electrons shoring through a very thin specimen, resulting in interactions between the electrons and the atoms allowing to observe particular features. The principle behind TEM is as same as light microscope, but uses electrons instead of light.

It can be used to study the growth of layers, their chemical composition as well as defects in semiconductors. High resolution can be used to analyze the quality, shape, size and density of nanoparticles/nanograins. Sharma et al. (2007) reported the intracellular accumulation of stable Au-NPs in roots and shoots of Sesbania drummondii seedlings.

TEM analysis of the root cells showed the presence of mono dispersed NPs in the organelles as well as multiple spherical NPs surrounding the cell organelles in the cytoplasm in the size range of 6-20 nm. Dubey et al. (2010) found that tansy fruit extract was used as a reducing agent in the synthesis of Au-NPs from auric acid and TEM images showed the formation of spherical and triangular NPs with an average size of 11 nm.

Wang et al. (2009) synthesized well dispersed NPs in the size range of 5-30 nm using the extract of Scutellaria barbata. Au-NPs synthesized by using the aqueous extract of Syzygium aromancum (clove) showed the formation of different morphologies such as triangular and polygonal shapes with an average size from 100 to 300 nm.

4.2   Carbon Nanotubes

Carbon Nanotubes can be Synthesized by any One of the Following Methods

- Pyrolysis.

- Laser evaporation.

- Carbons arc method.

- Chemical vapour deposition.

1. Pyrolysis

Carbon nanotubes are synthesized by the pyrolysis of hydro carbons such as acetylene at about 700°C in the presence of Fe-silica or Fe-graphite catalyst under inert conditions.

2. Laser Evaporation

A quartz tube containing argon gas and graphite target is heated to 1200°C. The quartz tube containing the water – cooled copper collector which is present outside the furnace. The graphite target containing small amount of cobalt and nickel which act as a catalyst for the formation of nanotubes.

An intense pulsed laser beam is incident on the target which results in evaporation of carbon from the graphite. The argon gas carried the carbon atoms from the high temperature zone to colder copper collector, where it condense into nanotubes.

3. Carbon Arc Method

It is carried out by applying direct current 60 – 100A and 20 – 25V are between graphite electrodes of 10 – 20μm diameter.

4. Chemical Vapour Deposition

It involves the decomposition of hydrocarbon gas (methane gas) at 1100°C. As the gas decomposes, carbon atoms are produced. Then it is condensed on a cooler substrate which contains catalyst like iron. This method produced tubes with open end which allows continuous fabrication. It is the most favorable method for scale up and production.

Carbon Nanotubes Properties and Applications

There are numerous CNT applications which take full advantage of CNTs unique properties of aspect ratio, mechanical strength, electrical and thermal conductivity.

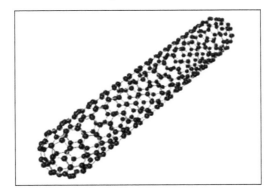

Carbon Nanotubes Properties

- They have High Electrical Conductivity.

- They have Very High Tensile Strength.

- They are highly Flexible- can be bent considerably without damage.

- They are Very Elastic ~18% elongation to failure.

- They have High Thermal Conductivity.

- They have a Low Thermal Expansion Coefficient.

- They are Good Electron Field Emitters.

- They have a High Aspect Ratio (length = ~ 1000 x diameter).

Carbon Nanotubes Applications

- Thermal Conductivity of CNT.

- Field Emission of CNT.

- Conductive Plastics with CNT.

- Energy Storage using CNT.

- Conductive Adhesives and Connectors with CNT.

- Molecular Electronics based on CNT.

- Thermal Materials with CNT.

- Structural Composites with CNT.

- Fibers and Fabrics with CNT.

- Catalyst Supports using CNT.

- Biomedical Applications of CNT.

- Air and Water Filtration using CNT.

- Ceramic Applications with CNT.

CNTs Electrical Conductivity

There has been considerable practical interest in the conductivity of CNTs. CNTs with particular combinations of N and M can be highly conducting and hence can be said to be metallic. CNTs can be either metallic or semi-conducting in their electrical behavior.

Conductivity in MWNTs is Quite Complex

Some types of "armchair"-structured CNTs appear to conduct better than other metallic CNTs. Additionally, interwall reactions within MWNTs have been found to redistribute the current over individual tubes non-uniformly.

Although, there is no change in current across different parts of metallic single-walled CNTs. However, the behavior of ropes of semiconducting SWNTs is different, in that the transport current changes abruptly at various positions on the CNTs.

The conductivity and resistivity of ropes of SWNTs has been measured by placing electrodes at different parts of the CNTs. The resistivity of the SWNT ropes was in the order of 10^{-4} ohm-cm at 27°C. This means that SWNT ropes are the most conductive carbon fibers known.

It has been reported that individual SWNTs may contain defects, these defects allow the SWNTs to act as transistors. Likewise, joining CNTs together may form transistor like devices. A nanotube with a natural junction behaves as a rectifying diode that is, a half-transistor in a single molecule. It has also recently been reported that the SWNTs can route electrical signals at high speeds when used as interconnects on semiconducting devices.

CNTs Strength and Elasticity

Carbon atoms of a single (graphene) sheet of graphite form a planar honeycomb lattice, in which each atom is connected via a strong chemical bond to three neighboring atoms. Because of these strong bonds, the basalplane elastic modulus of the graphite is one of the largest of any known material. For this reason, CNTs are expected to be the ultimate high-strength fibers.

The SWNTs are stiffer than steel and are very resistant to damage from physical forces. Pressing on tip of a nanotube will cause it to bend, but without damage to the tip. When the force is removed, the tip returns to its original state. This property makes CNTs very useful as probe tips for very high-resolution scanning probe microscopy.

Quantifying these effects has been rather difficult and an exact numerical value has not been agreed upon. Using an atomic force microscope (AFM), the unanchored ends of a freestanding nanotube can be pushed out of their equilibrium position and force required to push the nanotube can be measured.

The current modulus value of the SWNTs is about 1 TeraPascal, but this value has been disputed and a value as high as 1.8 Tpa has been reported. Other values significantly higher than that have also been reported.

The differences probably arise through different experimental measurement techniques. Others have shown theoretically that the modulus depends on the size and chirality of the SWNTs, ranging from 1.22 Tpa to 1.26 Tpa. They have calculated a value of 1.09 Tpa for a generic nanotube.

However, when working with different MWNTs, others have noted that the modulus measurements of MWNTs using AFM techniques do not strongly depend on the diameter. Instead, they argue that the modulus of the MWNTs correlates to the amount of disorder in nanotube walls. When MWNTs break, the outermost layers break first.

CNTs Thermal Conductivity and Expansion

New research from University of Pennsylvania indicates that CNTs may be the best heat conducting material man has ever known. Ultra-small SWNTs have even been shown to exhibit superconductivity below 20°K.

Research suggests that these exotic strands, already heralded for their unparalleled strength and unique ability to adopt the electrical properties of either semiconductors or perfect metals, may someday also find applications as miniature heat conduits in a host of the devices and materials. The strong in-plane graphitic C-C bonds make them exceptionally strong and stiff against axial strains.

The almost zero inplane thermal expansion but large interplane expansion of SWNTs implies strong inplane coupling and high flexibility against the non-axial strains. Many applications of CNTs, such as in the nanoscale molecular electronics, sensing and the actuating devices or as reinforcing additive fibers in functional composite materials, have been proposed.

The reports of several recent experiments on the preparation and mechanical characterization of CNT polymer composites have also appeared. These measurements suggest modest enhancements in strength characteristics of CNT embedded matrixes as compared to bare polymer matrixes.

The preliminary experiments and simulation studies on the thermal properties of CNTs show very high thermal conductivity. It is expected, therefore, that nanotube reinforcements in polymeric materials may also significantly improve thermal and thermo mechanical properties of the composites.

CNTs Field Emission

Field emission results from tunneling of electrons from a metal tip into vacuum, under application of a strong electric field. The small diameter and high aspect ratio of the CNTs is very favorable for field emission.

Even for moderate voltages, a strong electric field develops at the free end of supported CNTs because of their sharpness. This was observed by de Heer and co-workers at EPFL in 1995. He also immediately realized that these field emitters must be superior to conventional electron sources and might find their way into all kind of applications, most importantly flat panel displays.

Studying the field emission properties of MWNTs, Bonard and coworkers at EPFL observed that together with electrons light is emitted, as well. This luminescence is induced by electron field emission, since it is not detected without applied potential. This light emission occurs in visible part of the spectrum and can sometimes be seen with the naked eye.

CNTs High Aspect Ratio

The CNTs represent a very small, high aspect ratio conductive additive for plastics of all types. Their high aspect ratio means that a lower loading of CNTs is needed compared to other conductive additives to achieve the same electrical conductivity.

This low loading preserves more of the polymer resins' toughness, especially at low temperatures, as well as maintaining other key performance properties of the matrix resin.

CNTs have proven to be an excellent additive to impart electrical conductivity in plastics. Their high aspect ratio imparts electrical conductivity at lower loadings, compared to conventional additive materials such as carbon black, chopped carbon fiber or stainless steel fiber.

CNTs Conductive Plastics

Much of the history of plastics over the last half-century has involved their use as a replacement for the metals. For structural applications, plastics have made tremendous headway, but not where electrical conductivity is required, because the plastics are very good electrical insulators.

This deficiency is overcome by loading plastics up with conductive fillers, such as carbon black and larger graphite fibers. The loading required to provide the necessary conductivity using the conventional fillers is typically high, however, resulting in heavy parts and more importantly, the plastic parts whose structural properties are highly degraded.

It is well established that the higher the aspect ratio of filler particles, the lower the loading required needed to achieve a given level of conductivity. CNTs are ideal in this sense, since they have highest aspect ratio of any carbon fiber.

In addition, their natural tendency to form ropes provides inherently very long conductive pathways even at ultralow loadings.

Applications that exploit this behavior of the CNTs include EMI/RFI shielding composites coatings for enclosures, gaskets and other uses, electrostatic dissipation (ESD) and antistatic materials and conductive coatings and radar-absorbing materials for low-observable applications.

CNTs Energy Storage

They have the intrinsic characteristics desired in material used as electrodes in batteries and capacitors, two technologies of rapidly increasing importance. CNTs have a tremendously high surface area $(\sim 1000 \text{ m}^2/\text{g})$, good electrical conductivity and very importantly, their linear geometry makes their surface highly accessible to the electrolyte.

The Research has shown that the CNTs have the highest reversible capacity of any carbon material for use in lithiumion batteries. Furthermore, the CNTs are outstanding materials for super capacitor electrodes and are now being marketed for this application.

The CNTs also have applications in a variety of fuel cell components. They have a number of properties, including high surface area and thermal conductivity, which make them useful as electrode catalyst supports in PEM fuel cells.

They may also be used in gas diffusion layers, as well as current collectors, because of their high electrical conductivity. CNTs high strength and toughness to weight characteristics may also prove valuable as part of composite components in fuel cells that are deployed in the transport applications, where durability is extremely important.

CNTs Conductive Adhesives and Connectors

The same properties that make CNTs attractive as conductive fillers for use in electromagnetic shielding, ESD materials, etc., make them attractive for electronics packaging and interconnection applications, such as the adhesives, potting compounds and coaxial cables and other types of connectors.

CNTs Molecular Electronics

Building electronic circuits out of essential building blocks of materials molecules has seen a revival the past five years and is a key component of the nanotechnology. In any electronic circuit, particularly as dimensions shrink to the nanoscale, the interconnections between switches and other active devices become increasingly important.

Their geometry, electrical conductivity and the ability to be precisely derived, make CNTs the ideal candidates for the connections in molecular electronics. Furthermore, they have been demonstrated as switches themselves.

CNTs Thermal Materials

The record-setting anisotropic thermal conductivity of CNTs is enabling many applications where heat needs to move from one place to another place. Such an application is found in electronics, particularly advanced computing, where uncooled chips now routinely reach over 100°C.

The technology for creating aligned structures and ribbons of CNTs is a step toward realizing incredibly efficient heat conduits. In addition, composites with CNTs have been shown to dramatically increase their bulk thermal conductivity, even at very small loadings.

CNTs Structural Composites

The superior properties of CNTs are not limited to the electrical and thermal conductivities, but also include mechanical properties, such as stiffness, toughness and strength. These properties lead to a wealth of applications exploiting them, including advanced composites requiring high values of one or more of these properties.

CNTs Fibers and Fabrics

Fibers spun of pure CNTs have recently been demonstrated and are undergoing rapid development, along with CNT composite fibers. Such super strong fibers will have many applications including the body and vehicle armor, transmission line cables, woven fabrics and textiles. CNTs are also being used to make textiles stain resistant.

CNT Catalyst Supports

The CNTs intrinsically have an enormously high surface area in fact, for the SWNTs every atom is not just on a one surface each atom is on two surfaces, the inside and outside of nanotube. Combined with the ability to attach essentially any chemical species to their sidewalls provides an opportunity for unique catalyst supports. Their electrical conductivity may also be exploited in the search for new catalysts and catalytic behavior.

CNTs Biomedical Applications

The exploration of CNTs in biomedical applications is just underway, but has significant potential. Since a large part of the human body consists of carbon, it is generally though of as a very biocompatible material.

The cells have been shown to grow on CNTs, so they appear to have no toxic effect. The cells also do not adhere to CNTs, potentially giving rise to applications such as coatings for prosthetics, as well as anti-fouling coatings for ships.

The ability to functionalize the sidewalls of CNTs also leads to biomedical applications such as vascular stents and neuron growth and regeneration. It has a single strand of DNA can be bonded to a nanotube, which can then be successfully inserted into a cell.

CNTs Air and Water Filtration

Many researchers and corporations have already developed CNT based air and water filtration devices. It has been reported that these filters can not only block the smallest particles but also kill most bacteria.

CNTs Ceramic Applications

A ceramic material reinforced with the carbon nanotubes has been made by materials scientists at UC Davis. The new material is far tougher than conventional ceramics, conducts electricity and can both conduct heat and act as a thermal barrier, depending on the orientation of nanotubes.

The ceramic materials are very hard and resistant to heat and chemical attack, making them useful for applications such as coating turbine blades, but they are also very brittle. The researchers mixed powdered alumina with 5 to 10 percent carbon nanotubes and a further 5 percent finely milled niobium.

The researchers treated the mixture with an electrical pulse in a process called spark-plasma sintering. This process consolidates ceramic powders more quickly and at lower temperatures than conventional processes.

The new material has up to five times the fracture toughness resistance to cracking under stress of conventional alumina. The material shows electrical conductivity seven times that of previous ceramics made with nanotubes.

It also has interesting thermal properties, conducting heat in one direction, along the alignment of the nanotubes, but reflecting heat at right angles to the nanotubes, making it an attractive material for thermal barrier coatings.

Other Carbon Nanotubes Applications

There is a wealth of other potential applications for the CNTs, such as solar collection, nanoporous filters, catalyst supports and coatings of all sorts.

There are almost certainly many unanticipated applications for this remarkable material that will come to light in the years ahead and which may prove to be the most important and valuable ones of all.

Many researchers are looking into conductive and or water proof paper made with CNTs. The CNTs have also been shown to absorb Infrared light and may have applications in the I/R Optics Industry.

4.2.1 Carbon Fullerenes

Fullerenes are a family of carbon allotropes. molecules composed entirely of carbon in the form of a hollow sphere, ellipsoid, tube or plane. Fullerenes are similar in structure to graphite, which is composed of stacked sheets of linked hexagonal rings, but may also contain pentagonal or sometimes heptagonal rings.

The most familiar carbon fullerene is a molecule with 60 carbon atoms, represented as C_{60}. It was discovered in 1985 by Kroto et al. and named as Buckminsterfullerene. The name was coined after the American architect Richard Buckminster Fuller who was famous for the geodesic domes built by him.

The C_{60} molecule has a truncated icosahedral structure formed by replacing each vertex on the seams of a football by a carbon atom. There are 20 hexagonal faces and 12 pentagonal faces in the molecule.

The average nearest C - C distance is 0.144 nm, which is very close to that in graphite (viz. 0.142 nm). Each carbon atom is trigonally bonded to other carbon atoms, same as that in graphite. Out of the three bonds emanating from each carbon atom, there is one double bond and two single bonds.

The hexagonal faces consist of alternating single and double bonds and the pentagonal faces are defined by single bonds. The length of the single bonds is 0.146 nm, which is longer than the average bond length (i.e., 0.144 nm), while the double bonds are shorter, viz. 0.14 nm. The diameter of the C_{60} molecule is 0.71 nm. The inner cavity is capable of holding a variety of atoms.

Preparation of Fullerenes

A molecule composed of only carbon atoms in the form of a hollow sphere, ellipsoid or tube is called Fullerene. These are a new form of carbon, which were discovered in 1985 by the vapourisation of graphite under inert atmosphere at low pressure.

The properties of fullerenes are different from that of diamond and graphite. Fullerenes can be used in electronic devices, superconductors, catalysts, sensors, polymer composites, high energy fuels and medical applications. In graphite, we can find planar layer of carbon atoms, while in fullerenes, we can find tubular layers.

The spherical fullerenes are called buckyballs, while the cylindrical ones are called carbon nanotubes or buckytubes. The structure of the fullerene is similar to the structure of graphite, i.e., composed of stacked graphene an allotrope of carbon, which contains one-atom-thick planar sheets of densely packed sp^2-bonded carbon atoms.

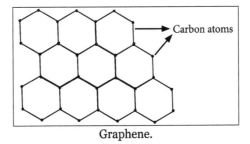

Graphene.

The name "graphene" is due to its resemblance with graphite and presence of sp^2 carbon atoms sheets of linked hexagonal rings, they may also contain pentagonal or heptagonal rings.

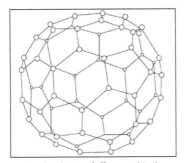

Buckminsterfullerene(C_{60}).

Buckminsterfullerene was the first discovered fullerene which possessed 60 atoms of carbon. It was prepared by Richard Smalley, Robert Curl, James Heath, Sean O'Brien and Harold Kroto at Rice University in 1985.

Properties

Chemical Properties

The carbon atoms within a Fullerene molecule are sp^2 and sp^3 hybridized, of which the sp^2 carbons are responsible for the considerably angle strain presented within the molecule. C_{60} and C_{70} exhibit the capacity to be reversibly reduced with up to six electrons.

This high electron affinity results from the presence of triply-degenerate low-lying LUMOs (lowest unoccupied molecular orbital). Oxidation of the molecule has also been observed, nevertheless, oxidation is irreversible. C_{60} has a localized pi-electron system, which prevents the molecule from displaying super aromaticity properties.

Physical Properties

Fullerenes are extremely strong molecules, able to resist great pressures—they will bounce back to their original shape after being subject to over 3,000 atmospheres. Theoretical calculations suggest that a single C_{60} molecule has an effective bulk modulus of 668 GPa when compressed to 75% its size.

This property makes fullerenes become harder than steel and diamond, whose bulk moduli are 160 GPa and 442 Gpa, respectively. An interesting experiment shows that Fullerenes can withstand collisions of up to 15,000 mph against stainless steel, merely bouncing back and keeping their shapes. This experiment resembles the high stability of the molecule.

Optical Properties

Delocalized pi electrons in Fullerenes are known to provide exceptionally large non-linear optical responses. Fullerenes have shown particular promises in optical limiting and intensity-dependent refractive index.

Additionally, the transfer of electrons from enclosed atom(s) to the Fullerene enhances the third-order nonlinear optical effect by orders of magnitude compared to empty cage Fullerenes.

Applications

Solar Cells

The high electron affinity and superior ability to transport charge make Fullerenes the best acceptor component currently available. First, they have an energetically deep-lying LUMO, which endows the molecule with a very high electron affinity relative to the numerous potential organic donors.

The LUMO of C_{60} also allows the molecule to be reversibly reduced with up to six electrons, thus illustrating its ability to stabilize negative charge. Importantly, a number of conjugated polymer–fullerene blends are known to exhibit ultrafast photoinduced charge transfer, with a back transfer that is orders of magnitude slower.

The state of the art in the field of organic photovoltaic is currently the Bulk Hetero junction (BHJ) solar cells based on Fullerene derivate phenylC_{61}butryric acid methyl ester (PCBM), whit reproducible efficiencies approaching 5 %.

Hydrogen Gas Storage

Due to its unique molecular structure, fullerene is the only form of carbon, which potentially can be chemically hydrogenated and de-hydrogenated reversibly. When fullerenes are hydrogenated, the C=C double bonds become CC single bonds and C-H bonds.

The bond strength of single C-C bonds is 83 kcal/mole and theoretical calculations show that the bond strength of the hydrogenated C-H bond is 68 kcal/mole. This means that for fullerene hydrides, the H-C bond is appreciably weaker that C-C bonds. Therefore, when heat is applied to fullerene hydrides, the H-C bonds will break before the C-C bonds and the fullerene structure should be preserved.

Fullerene Strengthening/Hardening of Metals

Fullerenes offer unique opportunities to harden metals and alloys without seriously compromising their ambient temperature ductility. This is due to the unique characteristics of fullerenes, namely their small size and high reactivity, which enable the dispersion strengthening of metallic matrices with carbide particles that result from insitu interactions between fullerenes and metals.

In a comparison of the hardness of a popular aerospace intermetallic compound Ti-24.5AI-17Nb, with and without fullerene additives, a 30% hardness enhancement was measured for the material with fullerene additives.

Fullerene as Precursor to Diamond

Fullerenes have proven to be an excellent precursor to diamond. This can be attributed in part to the curved structure of fullerenes, which possess a partial sp' (diamond) bonding configuration, as opposed to graphite, which is planar.

Argonne National Laboratories and MER Corporation have investigated fullerenes as the source of carbon to CVD deposit diamond with excellent results. MER has demonstrated excellent diamond coatings using fullerenes as the only source of carbon to deposit diamond at five times the growth rate of methane carbon sources, resulting in a projected 70% cost savings.

With low cost fullerenes it can be expected that conversion to diamond will be a very important commercial application.

Optical Application of Fullerene

Optical limiting refers to a decrease in transmittance of a material with increased incident light intensity. The phenomenon of optical limiting has a significant potential for applications in eye and sensor protection from intense sources of light.

Based on the optical limiting properties of fullerenes, one can make an optical limiter, which allows all light below an activation threshold to pass and maintains the transmitted light at a constant level below the damage threshold for the eye or the sensor.

4.3 Liquid Crystals: Types and Applications

Liquid crystals (LCs) are substances that exhibit long-range order in one or two dimensions, but not all three. Liquid crystals are substances that exhibit a phase of matter that has properties between those of a conventional liquid and those of a solid crystal.

A large number of organic molecules with long chain such as cholesteryl acetate $\left(CH_3COOC_{27}H_{45}\right)$, cholesteryl benzoate $\left(C_6H_5COOC_{27}H_{45}\right)$, etc. show LC behavior. A liquid crystal (LC) may flow like a ordinary liquid, but have the molecules in the liquid arranged and oriented in a crystal-like way or they show anisotropy and double refraction like the crystalline solids.

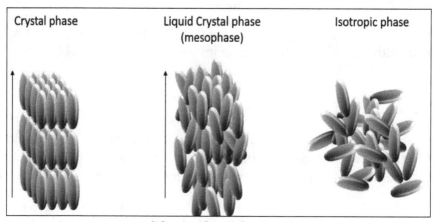

Solid - Liquid crystal – Liquid.

A liquid crystal is a thermodynamic stable phase characterized by anisotropy of the properties without the existence of a three-dimensional crystal lattice, generally lying in the temperature range between the anisotropic solid and isotropic liquid phase, hence the term mesophase.

Liquid crystal materials usually have several common characteristics. Among these are rod-like molecular structures, rigidness of the long axis and strong dipoles or easily polarizable substituents. The distinguishing characteristic of the liquid crystalline state is the tendency of the molecules to point along a common axis called the director. This is in contrast to molecules in the liquid phase, which have no intrinsic order.

In the solid state, molecules are highly ordered and has little translational freedom. The characteristic orientational order of the liquid crystal state is between the traditional solid and the liquid phases and this is the origin of the term mesogenic state, used synonymously with liquid crystal state.

It is sometimes difficult to determine whether a material is in a crystal or liquid crystal state. The crystalline materials the demonstrate long range periodic order in three dimensions.

Types

Liquid crystals can be classified into two main categories:

1. Thermotropic Liquid Crystals

It can be classified into two types: enantiotropic liquid crystals, which can be changed into liquid crystal state from either lowering the temperature of a liquid or raising of the temperature of a solid and the monotropic liquid crystals, which can only be changed into the liquid crystal state from either an increase in the temperature of a solid or a decrease in the temperature of a liquid, but not both.

In general, thermotropic mesophases occur because of anisotropic dispersion forces between the molecules and because of packing interactions.

2. Lyotropic Liquid Crystals

It occur with the influence of solvents, not by a change in temperature. Lyotropic mesophases occur as a result of solvent-induced aggregation of the constituent mesogens into micellar structures.

The Lyotropic mesogens are typically amphiphilic, meaning that they are composed of both lyophilic and lyophobic parts. This makes them to form into micellar structures in the presence of a solvent, since lyophobic ends will stay together as the lyophilic ends extend outward toward the solution.

As the concentration of the solution is increased and it is cooled, the micelles increase in size and eventually coalesce. This separates newly formed liquid crystalline state from the solvent.

Liquid Crystal Displays

Liquid crystals find wide use in liquid crystal displays, which rely on the optical properties of certain liquid crystalline molecules in the presence or absence of electric field. In a typical device, a liquid crystal layer sits between two polarizers that are crossed (oriented at 90° to one another).

The liquid crystal is chosen so that its relaxed phase is a twisted one. This twisted phase reorients light that has passed through the first polarizer, allowing it to be transmitted through second polarizer and reflected back to observer. The device thus appears clear.

When an electric field is applied to the LC layer, all the mesogens align. In this aligned state, mesogens do not reorient light, so the light polarized at the first polarizer is absorbed at the second polarizer and the entire device appears dark.

In this way, electric field can be used to make a pixel switch between clear or dark on command. The color LCD systems use the same technique, with color filters used to generate red, green and blue pixels. Similar principles can be used to make other liquid crystal based optical devices.

A liquid crystal display consists of an array of tiny segments that can be manipulated to present information. This basic idea is common to all displays, ranging from a full color LCD television to simple calculators.

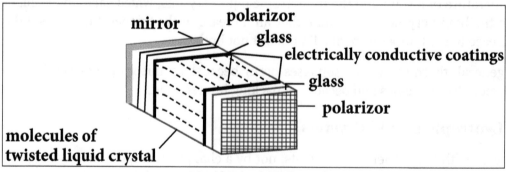

Liquid Crystal Displays.

Importance of Liquid Crystal Displays

The first factor is size. An LCD primarily consists of two glass plates with some liquid crystal material between them. There is no bulky picture tube. This makes LCDs practical for applications where sizes are important.

In general, LCDs use much less power than their cathode-ray tube (CRT) counterparts.

Many LCDs are reflective, meaning that they use only ambient light to illuminate the display. Even the displays that do require an external light source (i.e. computer displays) consume much less power than CRT devices.

Liquid crystal displays do have drawbacks; shorter lifetime of LC displays limits their use. Only AC drive can give a lifetime more than 10,000 hours whereas on DC excitation the maximum lifetime that can be obtained is 3000 hours.

Applications of Liquid Crystal

Liquid crystals have a multitude of other uses:

- They are used for nondestructive mechanical testing of materials under stress. This technique is also used for the visualization of RF (radio frequency) waves in wave-guides.

- They are used in medical applications where for example, transient pressure transmitted by a walking foot on the ground is measured.

- Low molar mass (LMM) liquid crystals have applications including erasable optical disks, full color "electronic slides" for computer-aided drawing (CAD) and light modulators for color electronic imaging.

- Some of the liquid crystals are used in hydraulic break/clutch system due to their high viscosity values.

- An application of liquid crystals that is now being explored is optical imaging and recording.

4.4   Super Conductors

The electrical resistance of metals depends upon temperature. Electrical resistance decreases with decreases in temperature and becomes almost zero near the absolute temperature. Materials in this state are said to possess super-conductivity. Thus super-conductivity may be defined as a phenomenon in which metals, alloys and chemical compounds become perfect conductors with zero resistivity at temperatures approaching absolute zero.

Super conductors are diamagnetic. The phenomenon was first discovered by Kammerlingh Onner in 1913 when he found that mercury becomes super conducting at 4K. The temperature at which a substance states behaving as super-conductor is called transition temperature. Which lies between 2 and 5K is most of the metals exhibiting this phenomenon.

Efforts are going on to find materials that behave as a super conductor at room temperature because attaining low temperature with liquid helium is highly expensive. The highest temperature at which super conductivity has been observed is 23K for alloys of niobium (Nb_3Ge) since 1987, many complex metal oxides have been found to possess super conductivity at fairly high temperatures. Some examples are given below.

$YBa_2Cu_3O_7 - 90K$

$Bi_2Ca_2Sr_2Cu_3O_{10} - 105K$

$TI_2Ca_2Ba_2Cu_3O_{10} - 125K$

Super conductors have many application in electronics, building magnets, aviation transportation (trains which move in air without rails) and power transmission.

Depending upon their behavior in an external magnetic field, super conductors are divided into two types:

- Type I superconductors.

- Type II superconductors.

Based on the value of H_c we have:

- Type I (or) Soft superconductors.

- Type II (or) Hard superconductors.

Type I Superconductor

In type I super conductor, the magnetic field is completely excluded from the material below the critical magnetic field and the material loses its superconducting property abruptly.

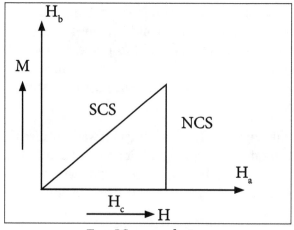

Type I Superconductor.

Characteristics

- They exhibit complete Meissner Effect.

- They have only one critical magnetic field value.

- Below, the material behaves as superconductor and above, the material behaves as normal conductor. These are called as Soft superconductors.

Type II Superconductor

In type II superconductor, the magnetic field is excluded from the material and the material loses its superconducting property gradually rather than abruptly.

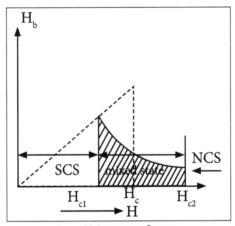

Type II Superconductor.

Characteristics

- They do not exhibit a complete Meissner Effect.

- They have two critical magnetic field values. Lower critical magnetic field $[H_{C_1}]$ and Higher critical magnetic field $[H_{C_2}]$.

- Below H_{C_1}, the material behaves as superconductor and above, the material behaves as normal conductor. The region in between $[H_{C_1}]$ and $[H_{C_2}]$ is called mixed state or vortex region. These are called as Hard superconductors.

Difference between Type I and II Superconductors

S. No.	Type I (or) Soft superconductors	Type II (or) Hard superconductors
1.	It exhibits complete Meissner Effect.	It does not exhibit a complete Meissner Effect.
2.	It is completely diamagnetic.	It is not completely diamagnetic.

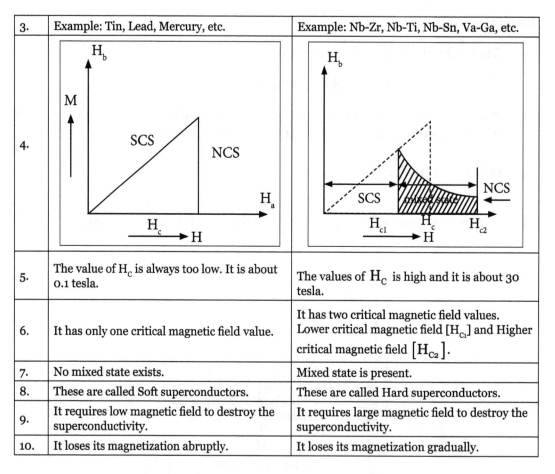

3.	Example: Tin, Lead, Mercury, etc.	Example: Nb-Zr, Nb-Ti, Nb-Sn, Va-Ga, etc.
4.		
5.	The value of H_c is always too low. It is about 0.1 tesla.	The values of H_C is high and it is about 30 tesla.
6.	It has only one critical magnetic field value.	It has two critical magnetic field values. Lower critical magnetic field $[H_{C_1}]$ and Higher critical magnetic field $[H_{C_2}]$.
7.	No mixed state exists.	Mixed state is present.
8.	These are called Soft superconductors.	These are called Hard superconductors.
9.	It requires low magnetic field to destroy the superconductivity.	It requires large magnetic field to destroy the superconductivity.
10.	It loses its magnetization abruptly.	It loses its magnetization gradually.

Characteristics of Superconductors

The properties of superconductors are as follows:

- The electrical resistance decreases by lowering the temperature and at transition temperature, the resistance abruptly becomes close to zero.

- The transition temperature is the characteristic of a particular isotope of an element.

- Transition temperature of superconductors decreases with increase in atomic mass of isotopes of an element.

- Presence of impurities in super conductors lowers their critical temperature (T_c).

- Below critical temperature, they conduct electricity without change of energy, i.e. $\Delta H = 0$. Therefore, current can flow through them for ever.

- All superconductors becomes diamagnetic and are repelled by magnets below critical temperature (T_c) and do not allow magnetic field to pass through it.

Applications of Superconductors

- In high speed computers called super computers for signal transmission, high memory storage and for other purposes.

- For making magnets for high energy particle accelerators.

- Used in magnetic resonance imaging (MRI) for medical diagnosis.

- In electric power transmission.

- For detecting very small direct current and voltages.

- In magnetically levitated trains using superconducting electromagnets.

4.5   Green Synthesis: Principles, 3 or 4 Methods of Synthesis and R4M4 Principles

Green synthesis approaches include mixed-valence polyoxometalates, polysaccharides, Tollens, biological and the irradiation method which have advantages over conventional methods involving chemical agents associated with the environmental toxicity.

Selection of solvent medium and selection of eco-friendly nontoxic reducing and stabilizing agents are the most important issues which must be considered in green synthesis of NPs.

Principles

1. Prevention

It is better to prevent waste than to treat or clean up waste after it has been created.

2. Atom Economy

The synthetic methods should be designed to maximize the incorporation of all materials used in the process into the final product.

3. Less Hazardous Chemical Syntheses

Wherever practicable, synthetic methods should be designed to use and generate substances that possess little or no toxicity to human health and the environment.

4. Designing Safer Chemicals

Chemical products should be designed to effect their desired function while minimizing their toxicity.

5. Safer Solvents and Auxiliaries

The use of the auxiliary substances (e.g., solvents, separation agents, etc.) should be made unnecessary wherever possible and innocuous when used.

6. Design for Energy Efficiency

Energy requirements of chemical processes should be recognized for their environmental and economic impacts and should be minimized. If possible, synthetic methods should be conducted at the ambient temperature and pressure.

7. Use of Renewable Feedstocks

The raw material or feedstock should be renewable rather than depleting whenever technically and economically practicable.

8. Reduce Derivatives

The unnecessary derivatization (use of blocking groups, protection/deprotection, temporary modification of physical/chemical processes) should be minimized or avoided if possible, because such steps require additional reagents and can generate waste.

9. Catalysis

Catalytic reagents (as selective as possible) are superior to stoichiometric reagents.

10. Design for Degradation

Chemical products should be designed so that at the end of their function they break down into innocuous degradation products and do not persist in the environment.

11. Real-time Analysis for Pollution Prevention

The analytical methodologies need to be further developed to allow for real-time, in-process monitoring and control prior to the formation of hazardous substances.

12. Inherently Safer Chemistry for Accident Prevention

The substances and the form of a substance used in a chemical process should be chosen to minimize the potential for chemical accidents, including releases, explosions and fires.

Physical Methods

Evaporation, condensation and laser ablation are the most important physical approaches. The absence of solvent contamination in the prepared thin films and the uniformity of NPs distribution are the advantages of physical synthesis methods in comparison with chemical processes.

The physical synthesis of silver NPs using a tube furnace at atmospheric pressure has some disadvantages, for example, the tube furnace occupies a large space, consumes a great amount of energy while raising the environmental temperature around source material and requires a lot of time to achieve thermal stability.

Moreover, a typical tube furnace requires power consumption of more than several kilowatts and a preheating time of several tens of minutes to reach a stable operating temperature. It was demonstrated that silver NPs could be synthesized via a small ceramic heater with a local heating area. The small ceramic heater was used to evaporate source materials.

The evaporated vapor can cool at a suitable rapid rate, because temperature gradient in the vicinity of the heater surface is very steep in comparison with that of a tube furnace. This makes possible the formation of small NPs in high concentration. The particle generation is very stable, because temperature of the heater surface does not fluctuate with time.

This physical method can be useful as a nanoparticle generator for long-term experiments for inhalation toxicity studies and as a calibration device for nanoparticle measurement equipment. The results showed that the geometric mean diameter, the geometric standard deviation and total number concentration of NPs increase with heater surface temperature.

Spherical NPs without agglomeration were observed, even at high concentration with high heater surface temperature. The geometric mean diameter and the geometric standard deviation of silver NPs were in the range of 6.2-21.5 nm and 1.23-1.88 nm, respectively.

Chemical Methods

Chemical Reduction

The most common approach for synthesis of silver NPs is chemical reduction by organic and inorganic reducing agents. In general, different reducing agents such as sodium citrate, ascorbate, sodium borohydride $(NaBH_4)$, elemental hydrogen, polyol process, Tollens reagent, N, N-dimethylformamide (DMF) and poly (ethylene glycol)-block copolymers are used for reduction of silver ions (Ag^+) in aqueous or non-aqueous solutions.

These reducing agents reduce Ag and lead to the formation of metallic silver (Ag^0), which is followed by the agglomeration into oligomeric clusters. These clusters eventually lead to the formation of metallic colloidal silver particles. It is important to use protective agents to stabilize the dispersive NPs during the course of metal nanoparticle preparation and protect the NPs that can be absorbed on or bind onto nanoparticle surfaces, avoiding their agglomeration.

The presence of surfactants comprising functionalities (e.g., thiols, amines, acids and alcohols) for interactions with particle surfaces can stabilize the particle growth and protect particles from sedimentation, agglomeration or losing their surface properties.

Electrochemical Synthetic Method

It can be used to synthesize silver NPs. It is possible to control particle size by adjusting the electrolysis parameters and to improve homogeneity of silver NPs by changing the composition of the electrolytic solutions.

Polyphenylpyrrole coated silver nanospheroids (3-20 nm) were synthesized by electrochemical reduction at the liquid/liquid interface. This nano-compound was prepared by transferring the silver metal ion from aqueous phase to organic phase, where it reacted with pyrrole monomer.

In another study, monodisperse silver nanospheroids (1-18 nm) were synthesized by the electrochemical reduction inside or outside zeolite crystals according to silver exchange degree of compact zeolite film modified electrodes. Additionally, spherical silver NPs (10-20 nm) with narrow size distributions were conveniently synthesized in aqueous solution by an electrochemical method.

Poly N-vinylpyrrolidone was chosen as stabilizer for the silver clusters in this study. Poly N-vinylpyrrolidone protects NPs from agglomeration, significantly reduces silver the deposition rate and promotes silver nucleation and silver particle formation rate.

Application of rotating platinum cathode effectively solves the technological difficulty of rapidly transferring metallic NPs from cathode vicinity to bulk solution, avoiding the occurrence of flocculates in vicinity of cathode and ensures monodispersity of particles. Addition of sodium dodecyl benzene sulfonate to the electrolyte improved particle size and particle size distribution of silver NPs.

Irradiation Methods

The Silver NPs can be synthesized by using a variety of irradiation methods. Laser irradiation of an aqueous solution of the silver salt and surfactant can produce silver NPs with a well-defined shape and size distribution. Furthermore, laser was used in a photo-sensitization synthetic method of making silver NPs using benzophenone.

At short irradiation times, low laser powers produced silver NPs of about 20 nm, while an increased irradiation power produced NPs of about 5 nm. The Laser and mercury lamp can be used as light sources for production of silver NPs. In visible light irradiation studies, photo-sensitized growth of silver NPs using thiophene (sensitizing dye) and silver nanoparticle formation by the illumination of $Ag(NH_3)^+$ in ethanol has been done.

Scheme of Synthesis

1. Scheme of Synthesis and Mechanism by conventional method.

2. Scheme of Synthesis and Mechanism by green chemistry method.

5

Water Technology

5.1 Hard Water and Alkalinity

Water is known as a natural solvent. Before it reaches the consumer's tap, it comes into contact with many different substances which includes organic and disease producing contaminants that may pollute the water.

Disinfection is one of the important step in the treatment of potable water, they are used to prevent diseases, can create byproducts which may pose significant health risks. Today, drinking water treatment at the point of use is no longer a luxury, it is a necessity.

Hardness

Hardness is defined as the concentrations of calcium and magnesium ions which is expressed in terms of calcium carbonate. These minerals in water can cause a few day to day problems. They react with soap to produce a deposit called soap curd that remains on the skin and clothes as it is insoluble and sticky, it cannot be removed by rinsing.

Hard water may also shorten the lifespan of plumbing and water heaters. When water containing calcium carbonate is heated, a hard scale is formed that can plug pipes and coat heating elements. Scale is a poor conductor of heat.

With increased deposits on the unit, heat does not get transmitted to the water fast enough and overheating of the metal results in its failure. Build-up of deposits will inturn reduce the efficiency of the heating unit, thereby increasing the cost of the fuel.

Reasons for Hardness

There are mainly three causes of hardness:

- Dissolved minerals.
- Dissolved Oxygen.
- Dissolved Carbon-di-oxide.

1. Dissolved Minerals

Dissolved minerals are of heavy metals. They get assimilated in water in the form of their soluble salts. Thus contact with minerals influences hardness. Dissolution is followed by hydration process in which minerals like $CaSO_4$ (Anhydrite) or Mg_2SiO_4 (Olivine) react with water as,

$$CaSO_4 + 2\ H_2O \xrightarrow{\text{Hydration}} CaSO_4.2H_2O$$

Anhydrous Gypsum (Volume increases by $\approx$ 33%),

$$Mg_2\ SiO_4 + xH_2O \xrightarrow{\text{Hydration}} Mg_2\ SiO_4.xH_2O$$

Olivine Serpentine

2. Dissolved Oxygen

Dissolved oxygen is also one of the major factor to influence the hardness in water. D.O. influences oxidation and hydration of metal oxides/sulphides as,

$2\ Fe_3O_4$	$+ O_2$	$3Fe_2O_3$	$+ 2\ H_2O$	$3Fe_2O_3 \cdot 2H_2$
Magnetite	Oxidation	Haematite	Hydration	OLimonite

$2FeS_2 + 7O_2 + 2H_2O$	$2FeSO_4 + 2H_2SO_4$
Mercesite	

3. Dissolved CO_2

The pH of water decreases due to dissolution of CO_2 from atmosphere. Due to this the dissolution of other minerals also increases.

The following reactions are self-explanatory that how dissolved CO_2 influences hardness of water,

(i) $CaCO_3 + CO_2 + H_2O \rightarrow Ca\left(HCO_3\right)_2$

(Insoluble) from rock Soluble

(ii) $MgCO_3 + CO_2 + H_2O \rightarrow Mg\left(HCO_3\right)_2$

(Insoluble) from rock Soluble

(iii) CO_2 also influences conversion of Ca / Na / K / Fe silicates and aluminium silicates (present in rocks) into soluble carbonates and bicarbonates. Some of them get converted into free silica. e.g.

$$K_2O.\ Al_2O_3\ .\ 6SiO_2 + CO_2 + 2H_2O \rightarrow Al_2O_3\ .\ 2SiO_2\ .\ 2H_2O + K_2CO_3 + 4SiO_2$$

These products i.e. dissolved salts, fine clay and free silica get collected in water, increasing its hardness.

Units for Measuring Hardness

- mg/liter of $CaCO_3$.

- Parts per million of $CaCO_3$.

Usually, the hardness of water is expressed in terms of calcium carbonate equivalents. The formula used to convert the mass of hardness producing salt to mass of $CaCO_3$ equivalents is given below:

$$\text{Calcium carbonate equivalents} = \frac{\text{Mass of salt} \times \text{Molecular mass of } CaCO_3}{\text{Molecular mass of salt}}$$

Note: Molecular masses of hardness producing salts are given below:

Hardness producing salt	Molecular Mass
$CaSO_4$	136
$MgSO_4$	120
$CaCl_2$	111
$MgCl_2$	95
$Ca(HCO)$	162
$Mg(HCO_3)_2$	146
$CaCO_3$	100

There are two types of water hardness:

- Temporary

- Permanent

Permanent hardness in water is the hardness due to the presence of the chlorides, nitrates and sulphates of calcium and magnesium, which will not be precipitated by boiling. The lime scale can build up on the inside of the pipe restricting the flow of water or causing a blockage. This can happen in the industries where hot water is used.

Temporary Hardness is due to the bicarbonate ion, HCO_3, being present in the water. This type of hardness can be removed by boiling the water to expel the CO_2, as indicated in the equation below:

$$Ca(HCO_3)_2 \leftrightarrow CaCO_3 + CO_2 + H_2O$$

Permanent hardness is due to presence of calcium and magnesium nitrates, sulphates and chlorides etc. This type of hardness cannot be eliminated by the method of boiling.

Temporary Hardness is due to the bicarbonate ion, HCO_3^-, being present in the water. It can be removed by water reboiling, whereby white solid emerges calcium carbonate that is limescale.

$$Ca(HCO_3)_2 \leftrightarrow CaCO_3 + CO_2 + H_2O$$

Health Effects of Hardness

The presence or absence of the hardness minerals in drinking water is not known to cause a health risk to the consumers. Hardness is generally considered an aesthetic water quality factor. The presence of some dissolved mineral material in drinking water is the reason for the water characteristic and pleasant taste.

At higher concentrations however, hardness can create the following consumer issues:

- Produces a white mineral deposit on the dishes noticeable on clear glassware.

- Reduces the efficiency of the devices that helps in heating of water. As hardness deposits integrate thickness, they act like insulation, reducing the efficiency of the heat transfer.

- Produces soap scum which is most noticeable on tubs and showers.

- It has also been determined that the areas of higher hardness in drinking water may be related to lower incidents of heart disease. This potential relationship is now being investigated.

Disadvantages of Hard Water

- Scales have a thermal conductivity value which is less than that of the bare steel. Even a very thin layer of scales serves as an insulator and resist in the transfer of heat.

- Internal diameter of the pipes of the boilers progressively decreases due to the formation of scales.

- Evaporative capacity of the boiler surface reduces due to the scale formation as they block the passage of heat transfer.

- Surface of tubes becomes rough and resist the proper flow of the water.

- Different parts of the boiler become weak and distorted due to overheating. As a result, operation of boilers becomes dangerous especially in the high pressure boilers.

- Boiler efficiency decreases as the valves and condensers of the boilers are choked.

- Due to scale formation boiler tubes are clogged down.

- Water does not come in direct contact with tubes and plates of boilers due to the formation of scales. This results in overheating and sometimes burning of these plates and tubes.

1. Domestic uses

- Washing

- Bathing

- Drinking

- Cooking

$$C_{17}H_{35}COO\,Na + H_2O \rightarrow C_{17}H_{35}COOH + NAOH$$

$$C_{17}H_{35}COOH + C_{17}H_{35}COO\,Na \rightarrow Lather$$

$$C_{17}H_{35}COO\,Na + CaSO_4 \rightarrow \left(C_{17}H_{35}COO\right)_2 Ca \downarrow + Na_2SO_2$$

2. Industrial uses

- Boiler Feed: Should not contain nitrates- scale and sludges.

- Paper Mill: Should not contain iron and lime- destroy resin of soap.

- Sugar industries: Sulphates and Alkaline carbonates- Deliquescent.

- Dyeing Industries: Should not contain iron and hardness.

- Laundries: Should be soft.

Determination of Hardness

EDTA Method

Ethylene diamine tetra acetic acid (EDTA) is a reagent that helps in the formation of EDTA-metal complexes with many metal ions. In alkaline conditions (pH>9) it forms stable complexes with the alkaline earth metal ions Ca^{2+} and Mg^{2+}.

The EDTA reagent can be used to measure the total quantity of the dissolved Ca^{2+} and Mg^{2+} ions present in a water sample. Thus the total hardness of the water sample can be estimated by the method of titration with a standard solution of EDTA.

Suitable conditions for the titration are achieved by the addition of a buffer solution having a pH value of 10. The buffer solution stabilizes the pH at 10. There are H+ ions produced as the reaction proceeds and without the buffer solution the pH would decrease.

The EDTA reagent cannot under these conditions distinguish between the hardness caused by Ca^{2+} and Mg^{2+} or (directly) between temporary and permanent hardness. Therefore the results of this experiment are generally expressed in terms of the quantity of insoluble $CaCO_3$ that would have to be converted into soluble salts to give the same total number of moles of dissolved Ca^{2+} and Mg^{2+} ions. This enables the total hardness of water from various sources to be compared easily.

Because it is a primary standard and is also more soluble in water, the di sodium salt of the EDTA is more commonly used as a reagent rather than the EDTA itself. If Na_2H_2Y represents the above salt, it ionizes in the aqueous solution to form H_2Y^{2-}, which complexes in a 1:1 ratio with either Ca^{2+} or Mg^{2+} ions (which are represented as M^{2+}). The reaction can be represented as follows:

$$H_2Y^{2-} + M^{2-} \rightarrow MY^{2-} + 2H^+$$

The indicator Eriochrome Black T is used to detect the end point. This is an indicator that has a different colour when complexed to the metal ions than when it is a free indicator.

The reaction between the red indicator-metal complex and the ETDA reagent at the end point can be represented as follows:

$$MIn^- + H_2Y^{2-} \rightarrow HIn^{2-} + MY^{2-} + H^+$$

Procedure

- Wash the pipette, burette and conical flask with the deionized water. Rinse the burette with the EDTA solution and the pipette with the hard water.

- Using the funnel, fill the burette with EDTA solution. Open the tap to fill the part below the tap. Then remove the funnel. Adjust the level of the solution to zero mark. Make sure that the burette is vertical.

- Use the pipette to transfer 50 cm³ of the hard water sample to the conical flask. Add 2-3 cm³ of the buffer (pH 10) solution.

- Add 0.03 g of the solid indicator to the contents of the flask in the following manner: Add gradually to the flask, swirling after each addition. A deep wine red colour is obtained.

- Carry out one 'rough' titration to find an approximate end point, followed by a number of accurate titrations until two titres agree to within 0.1 cm³. At the end point, the colour should be dark blue, with no tinge of wine-red colour.

- From the data, calculate the total hardness of the water sample.

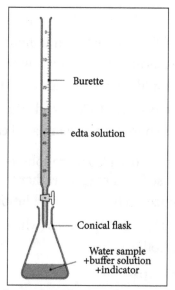

Experimental verification of EDTA solution.

Advantages of EDTA Method

- EDTA method shows the result with greater accuracy.

- This method is more convenient in comparison with other methods.

- Procedure of EDTA method is more rapid.

Alkalinity

Alkalinity is a chemical measurement of a water's ability to neutralize acids. Alkalinity is also a measure of a water's buffering capacity or its ability to resist changes in pH upon the addition of acids or bases. Alkalinity of natural water is due to the presence of weak acid salts although strong bases may also contribute (i.e. OH⁻) in extreme environments.

Bicarbonates represent the major form of alkalinity in natural water and its source being the partitioning of CO_2 from the atmosphere and the weathering of carbonate minerals in rocks and soil. Other salts of weak acids, such as borate, silicates, ammonia, phosphates and organic bases from natural organic matter, may be present in small amounts. Alkalinity, by convention, is reported as mg/L $CaCO_3$ since most alkalinity is derived from the weathering of carbonate minerals.

Parameters of Alkalinity

- The alkalinity of a body of water provides information about how sensitive that water body will be to acid inputs such as acid rain.

- Turbidity is frequently removed from drinking water by coagulation and flocculation. This process releases H^+ into the water. Alkalinity must be present in excess of that destroyed by the H^+ released for effective and complete coagulation to occur.

- Hard water are frequently softened by precipitation methods. The alkalinity of the water must be known in order to calculate the lime $\left(Ca(OH)_2\right)$ and soda ash $\left(Na_2CO_3\right)$ requirements for precipitation.

- Alkalinity is important to control corrosion in piping systems.

Bicarbonate $\left(HCO^{3-}\right)$ and carbonate $\left(CO_3^{2-}\right)$ may combine with other elements and compounds, altering their toxicity, transport and fate in the environment.

Three Types of Alkalinity

Alkalinities are classified according to the endpoint of titration with strong acid:

- M alkalinity = Alk = Total alkalinity (endpoint: CO_2 equivalence point).

- P alkalinity = Carbonate alkalinity (endpoint: HCO^{3-} equivalence point).

- Caustic alkalinity (endpoint: CO_3^{-2} equivalence point).

Here, "M" refers to the pH indicator methyl orange (endpoint 4.2 to 4.5), "P" refers to the pH indicator phenolphthalein (endpoint 8.2 to 8.3). "M alkalinity" is usually call "alkalinity" or "general alkalinity" or "total alkalinity".

For a simple carbonate system alkalinities are defined by:

- M alkalinity $=\left(\left[OH^-\right]-\left[H^+\right]\right)+\left[HCO_3^-\right]+2\left[CO_3^{-2}\right]$

- P alkalinity $= \left(\left[OH^- \right] - \left[H^+ \right] \right) + \left[CO_3^{-2} \right] - \left[H_2CO_3^* \right]$

- Caustic alkalinity $= \left(\left[OH^- \right] - \left[H^+ \right] \right) - \left[HCO_3^- \right] - 2 \left[H_2CO_3^* \right]$

Substracting M alkalinity from P alkalinity yields the total amount of dissolved inorganic carbon:

$$M - P = \left[H_2CO_3^* \right] + \left[HCO_3^- \right] + \left[CO_3^{-2} \right] = DIC$$

Where the asterisk on $H_2CO_3^*$ symbolizes the composite carbonic acid.

The equations above are restricted to carbonate systems without other weak acids or bases. The formula for the general case is given here, which is used in standard hydro chemistry programs. The calculated values are presented in output tables.

Phenolphthalein and Methyl Orange Alkalinity

Alkalinity is defined as a measure of the buffering capacity of water to neutralize strong acid. This capacity is attributed to bases that are present in natural waters including OH^-, HCO_3^- and CO_3^{2-}. More alkalinity in water sample means more buffering capacity of water sample.

In order to determine Alkalinity of samples, samples will be titrated with sulfuric acid or hydrochloric acid to a certain pH end point (pH 8.3 for Phenolphthalein Alkalinity and pH 4.5 for Total Alkalinity) and the volume of the acid used for the titration will be recorded.

The following reactions are occurring during titration:

pH range above 8.3

$$OH^- + H^+ \rightarrow H_2O$$

$$CO_3^{2-} + H^+ \rightarrow HCO_3^-$$

pH range between 8.3 and 4.5

$$HCO_3^- + H^+ \rightarrow H_2CO_3$$

If carbonate $\left(CO_3^{2-} \right)$ is present in sample, CO_3^{2-} will consume one H^+ when the solution is titrated to pH 8.3 and it will consume another H^+ during further titration from pH 8.3 to pH 4.5. If the volume of acid required to get pH 8.3 is equal to the volume of acid used from pH 8.3 to pH 4.5, sample contained only CO_3^{2-} (no OH^-) as the major alkalinity component.

If the pH of sample is below 8.3 and a certain amount of acid is required to reach pH

4.5, then sample contains only HCO_3^- (no OH^- and CO_3^{2-}). If sample requires a certain amount of acid to reach pH 8.3, but no acid is required from pH 8.3 to pH 4.5, then sample contains only OH^-.

Calculation

Phenolphthalein Alkalinity = Amount of acid used to reach pH 8.3 (ml) * Normality of acid (eq/L) x 100,000 $\left(mg\ CaCO_3 / eq \right)$ / sample volume (ml).

Where Normality = Molarity (moles/L) x the number of hydrogen exchanged in a reaction (eq/moles).

Total Alkalinity = Amount of acid used to reach pH 4.5 (ml) x Normality of acid (eq/L) x 50,000 $\left(mg\ CaCO_3 / eq \right)$ / sample volume (ml).

Procedure

- Measure 50 ml or 100ml of sample into a 250 mL beaker or erlenmeyer flask. Place sample onto a stir plate (make sure to put a bar magnet in the flask).

- Measure initial pH of sample. If the sample pH is below 8.3 (if above 8.3, do step 3 first), add several drops of bromcresol green indicator. If the color of the solution turns blue, titrate sample with 0.02 N H_2SO_4 or HCl (we need to dilute the acid provided) until the color changes to yellow Record the total volume of acid used for the titration.

- Measure initial pH of sample. If the sample pH is above 8.3, add several drops of phenolphthalein indicator. If the color of the solution turned pink, titrate sample with 0.02 N H_2SO_4 or HCl (we may need to dilute the acid provided in the lab) until color changes from pink to clear (pH 8.3). Record the volume of acid used for the titration. Then, proceed with step 2.

- Calculate both Phenolphthalein Alkalinity and Total Alkalinity using the formula provided above.

5.2   Water for Steam Generation and Boiler Troubles

In industry, one of the chief uses of water is generation of steam by boilers. Water that is fed into the boiler for the production of steam is called as boiler feed water.

Requirements of a Boiler Feed Water

- It should be free from dissolved gases, silica, suspended matter and oil.

- It should be free from boiler troubles like priming, foaming, scale and sludge.

- The water should have hardness less than 0.2 ppm.

- Alkalinity should be less than 1 ppm.

Boiler Troubles

Water is used in boilers, steam engines etc., to raise steam. When a sample of hard water is used in boiler to prepare steam, the following problems will occur.

- Priming and foaming.

- Scale formation.

- Corrosion of boiler metal.

- Caustic Embrittlement.

1. Priming and Foaming

Priming

When the steam is produced in the boiler, due to rapid boiling, some droplets of liquid water are carried along with steam. Steam containing droplets of liquid water is called wet steam. These droplets of liquid water carry with them some dissolved salts and suspended impurities. This phenomenon is called carry over. It occurs due to priming and foaming.

Causes

- Sudden boiling.

- Very high water level in boiler.

- Presence of excessive foam on the surface of the water.

- Poor boiler design.

Prevention of Priming

- Good boiler design.

- Providing proper evaporation of water.

- Maintaining uniform heat distribution.

- Adequate heating surface.

- Maintaining low water levels.

Foaming

The formation of stable bubbles above the surface of water is called foaming. These bubbles are carried over by steam leading to excessive priming.

Causes

- Presence of oil and grease.
- Presence of finely divided particles.

Prevention of Foaming

- Adding anti foaming agents like castor oil and small amount of polyamide.
- Clay, suspended solids, droplets of oil and grease can be removed by treated water with clarifying agents such as hydrous silicic acid and aluminium hydroxide.
- The concentration of salts and sludge in the boiler can be controlled by internal treatment and blow down operation.

2. Scale Formation

Scale

Scale formation in boiler is due to the presence of various salts which comes out of solution and deposited because of the effects of temperature and density. The salts of calcium and magnesium are the major sources of scale formation.

Scale is hard, thick, strong adherent precipitate due to salts like $CaSO_4$, $Ca(HCO_3)_2$. It can be prevented by special methods like:

- External treatment of ion exchange.
- Internal carbonate, phosphate.
- Mechanical hard scrubbing methods.

Typical constituents of scale and deposits in boiler are:

- Calcium Carbonate: $CaCO_3$.
- Calcium Sulphate: $CaSO_4$.
- Calcium Phosphate: $Ca_3(PO_4)_2$.
- Magnesium Hydroxide: $Mg(OH)_2$.
- Magnesium Phosphate: $Mg_3(PO_4)_2$.

- Iron and Copper Oxides: Fe_2O_3, CuO.

- Complex silicates of Magnesium, Iron, Sodium and aluminium.

Sludge

If the precipitate is loose and slimy it is called sludges. Sludges are formed by substances like $MgCl_2$, $MgCO_3$, $MgSO_4$ and $CaCl_2$. They have greater solubilities in hot water than cold water.

Disadvantages of Scale Formation

- Scales have a thermal conductivity value less than the bare steel. Even a very thin layer of scales serves as an insulator and resist in heat transfer.

- Internal diameter of the pipes of boilers progressively decreased due to scale formation.

- Evaporative capacity of the boiler surface was reduced due to scale formation as they blocked the passage of heat transfer.

- Surface of tubes becomes rough and resist to proper flow of the water.

- Different parts of the boiler become weak and distorted due to overheating. As a result, operation of boilers becomes dangerous especially in high pressure boilers.

- Boiler efficiency decreases because the valves and condensers of boilers are choked.

- Due to scale formation boiler tubes clog down.

- Water does not come in direct contact with tubes and plates of boilers due to scale formation. This results in overheating and sometimes burning out of these plates and tubes.

Boiler Explosion

When thick scales crack due to uneven expansion, the water comes out suddenly in contact with overheated iron plates. This causes formation of a large amount of steam suddenly. So sudden high pressure is developed which may even cause the explosion of the boiler.

Causes for Sludge and Scale

- The salt deposit formed is a poor conductor of heat. Therefore fuel is wasted in raising the temperature of the boiler.

- Due to the increase in the temperature, the plates may melt. This may lead to explosion of boiler.

- At higher temperature, more oxygen may be absorbed by the boiler metal which causes corrosion of boiler metal.

- The sudden spalling of the boiler scale exposes the hot metal suddenly to super-heated steam which causes corrosion of boiler.

Prevent the sludge and scale formation by Internal conditioning method and External conditioning methods:

External Treatment

- Zeolite process.

- Lime soda process.

- Ion exchange process.

Internal Treatment

- Colloidal conditioning.

- Phosphate conditioning.

- Carbonate conditioning.

- Calgon conditioning.

- Treatment with sodium aluminate.

3. Boiler corrosion

Boiler corrosion occurs due to dissolved gases such as oxygen, CO_2, SO_2 and H_2S, due to hydrolysis of dissolved salts such as $MgCl_2$ in the boiler water. As a result deep holes are formed in the boiler, these can be minimized by using the following methods.

Prevention of Oxygen by Chemical Method

i) Adding Sodium Sulphite:

$$Na_2SO_3 + O_2 \rightarrow 2Na_2SO_4$$

This method results in other precipitates which can have some side effects. So this method is less preferred.

ii) Adding Hydrazine:

$$N_2H_4 + O_2 \rightarrow N_2 \uparrow + 2H_2O$$

This method results in inert gas and pure water and has no side effects. So it is preferred.

Prevention from CO_2

- Chemical method: By adding calculated amount of ammonium hydroxide.

$$2NH_4OH + CO_2 \rightarrow \left(NH_4\right)_2 CO_3 + H_2O$$

- Mechanical de-aeration method.

- Corrosion due to dissolved salts like $MgCl_2$ by Neutralization: Excess acidic nature is neutralized by adding alkalis and vice versa.

$$HCl + NaOH \rightarrow NaCl + H_2O$$

Types of Corrosion

There are different kinds of corrosion which depend on the environment surrounding the material, type of material, chemical reaction etc. Some common types of corrosion are described below:

i. Uniform Corrosion

This is also known as General corrosion. It is a very general method of corrosion. It deteriorates the whole surface of the metal and makes the surface thin. The damage is done at a constant rate on the entire surface. It will be easily detected by it's appearance. It will be controlled but if it is not, it then destroys the whole metal.

ii. Galvanic Corrosion

This kind of corrosion happens with an electrolyte like seawater. Metals have different values of electrical potentials. When they are electrically connected, put in an electrolyte, the more active metal which has a high negative potential becomes the anode. Since it has high negative potential, it corrodes fast. But the less active metal becomes the cathode.

Therefore the flow of electric current continues till the potentials are equal between both electrodes. Thus, at joints where the two non-similar metals meet, the galvanic corrosion appears. The Galvanic Series shows the list of metals from most active to least active.

Thus galvanic corrosion will be controlled by selecting the two metals which are close in series. Since the platinum is the least active, it is also less active for corrosion.

iii. Pitting Corrosion

This happens because of random attacks on particular parts of the metal's surface. This makes holes which are large in depth. These holes are known as "pits".

The pit acts as the anode while the undamaged part of the metal is the cathode. It starts with a chemical breakdown in the form of a scratch or spot. The pitting process makes the metal thinner and increases fatigue. Example, it will be very harmful in gas lines.

iv. Stress Corrosion Cracking (SCC)

SCC is a complex form of corrosion which arises due to stress and corrosive environment. This generates brittle and dry cracks in the material. The stress is developed in the material is due to the bending or stretching of the material. It additionally affects only at a particular section of material.

Specific reasons for stress corrosion are welding, heating treatments, deformation etc. It is very difficult to detect the cracks or to detect the stress corrosion because they combine with active path corrosion. The active path corrosion occurs normally along grain or crystallographic boundaries.

Stress corrosion is strongly affected by alloy composition.

v. Corrosion Fatigue

This happens in the presence of a corrosive environment like saltwater. It is a combination of corrosion and cyclic stress. The corrosion fatigue is produced when a metal breaks at a stress level which is lower than its tensile strength. This is strongly affected by the environment in which the metal resides which affects the initiation and growth rate of the cracks.

Those cracks are too fine to detect easily. Therefore the stress coupons are used to detect the corrosion. This will be produced by the influence of various types of stress like stresses applied, thermal expansion, welding, soldering, cleaning, thermal contraction, heating treatment, construction process, casting etc.

To prevent the corrosion fatigue, the designing and construction process of the materials must be done properly, by eliminating any stress and the environmental factors and by eliminating the crevices.

vi. Intergranular Corrosion

In the granular composition of metals and alloys, grains are present and their surfaces join with each other. This forms the grain boundaries. Thus the grains are separated by grain boundaries. The intergranular corrosion is also called inter crystalline corrosion.

The intergranular corrosion is being developed on or near the grain boundaries of a metal. This will be due to welding, stress, heat treating or improper service etc. The metal will loose its strength due to the Intergranular corrosion.

vii. Crevice Corrosion

It is also called concentration cell corrosion. This is due to the trapping of liquid corrosive between the gaps of the metal. Since the electrolyte has aggressive ions like chlorides, the corrosion reaction is started after settling of liquid in gaps.

The oxygen is consumed during the reaction. Thus an anodic area is developed near the oxygen-depleted zone where the external part of the material acts as a cathode. Crevice corrosion is similar to pitting corrosion. It is very difficult to detect crevice corrosion. It will be initiated by materials like gaskets, fasteners, surface deposits, washers, threads, clamp etc.

viii. Filiform Corrosion

It is a kind of concentration cell corrosion. This develops on coated metallic surfaces with a thin organic film. The corrosion generates the defect on the protective coating of metallic surface. The filaments of corrosion product is the cause of degradation of the coating.

The filaments look like thin threads. The actively growing filaments do not intersect the inactive filaments. They exist as long branching paths. The reflection process takes place when filaments collide with each other. Filiform corrosion is a very specific process since it only affects the surface's appearance, not the metallic material.

ix. Erosion Corrosion

It is also known as flow-assisted corrosion. This is due to the movement of corrosive liquids on metal surface which damages the material. It will be seen in ship propellers which are constantly exposed to sea water or in soft alloys.

The damage will be seen as waves or rounded holes etc. This shows the flow of the corrosive liquid. It will be controlled by the use of hard alloys, then managing the velocity and flow pattern of the fluid.

x. Fretting Corrosion

It is a form of erosion-corrosion. Due to this corrosion, the material surface starts to disappear. The fretting corrosion exists in the form of dislocations of the surface and deep pits. Oxidation is the main cause of fretting corrosion. It will be controlled by using lubricates, controlling movement etc.

4. Caustic Embrittlement

It is a phenomenon that occurs in boilers where caustic substances (NaOH) accumulate in boiler materials. Residual sodium carbonate used for the softening process undergoes hydrolysis forming sodium hydroxide at high pressure and temperatures.

The alkaline water enters the minute holes and cracks by capillary action on the interior boiler. The water then diffuses out of the cracks and leaving behind hydroxide salts that accumulate when more water evaporates.

The hydroxide attacks the surrounding material of the boiler and dissolves iron as sodium ferrite.

$$Na_2CO_3 + H_2O \rightarrow 2NaOH + CO_2^-$$

$$3Fe + 3NaOH \rightarrow 3Na_2FeO_2 \ (sodium \ ferroate)$$

$$3Na_2FeO_2 + 4H_2O \rightarrow 6NaOH + Fe_3O_4 + H2$$

Further dissolution of iron depends on regeneration of sodium hydroxide.

Prevention of Caustic Embrittlement

- As softening agent, we can use sodium phosphate instead of sodium carbonate.
- The hair line cracks can be sealed by waxy materials like Tannin and Lignin.

5.3 Internal Treatments

The residual salts that are not removed by external methods can be removed by directly adding some chemicals into the boiler water. These chemicals are called as Boiler compounds and the method is known as 'Internal treatment'.

The various examples are Carbonate conditioning, Phosphate conditioning, Calgon conditioning, etc.

1. Carbonate Conditioning

Used for The low pressure boilers. Here the salts like $CaSO_4$ are converted to easily removable $CaCO_3$. But some times it produces NaOH, CO_2 and hence carbonic acid is produced. So it is less preferred.

$$CaSO_4 + Na_2CO_3 \rightarrow CaCO_3 + Na_2SO_4$$

2. Phosphate Conditioning

Used for high pressure boiler. No such risk of CO_2 liberation.

$$3CaSO_4 + 2Na_3PO_4 \rightarrow Ca_3(PO_4)_2 + 3Na_2SO_4$$

3. Calgon Conditioning

Calgon is the trade name of sodium hexa meta phosphate- $Na_2\left[Na_4(PO_3)_6\right]$. With calcium ions it forms a soluble complex and prevents scale and sludge formation. It is used for high and low pressure boilers.

$$2CaSO_4 + Na_2\left[Na_4(PO_3)_6\right] \rightarrow Na_2\left[Ca_2(PO_3)_6\right] + 2Na_2SO_4$$

4. Colloidal Conditioning

In case of low pressure boilers, scale formation can be avoided by adding organic substances like Kerosene, tannin, agar-Agar etc. which get coated over the scale forming precipitates Which yields coated non sticky and loose deposits.

5. Treatment with Sodium Aluminate ($NaAlO_2$)

Sodium aluminate gets hydrolyzed yielding NaOH and a gelatinous precipitate of aluminium hydroxide.

$$NaAlO_2 + 2H_2O \rightarrow NaOH + Al[OH]_2$$

The NaOH so formed precipitate some of the magnesium as $Mg(OH)_2$ i.e.,

$$MgCl_2 + 2\,NaOH \rightarrow Mg(OH)_2 + 2NaCl$$

The precipitate of $Mg(OH)_2$ and $Al(OH)_3$ produced inside the boiler entraps finely suspended and Colloidal impurities including oil drops and silica.

Chemicals used in the Internal Treatment

Phosphates were used as the main conditioning chemical, however nowadays chelate and polymer type chemicals are mostly used. These new chemicals have the advantage over the phosphates of maintaining a scale-free metal surfaces.

All internal treatment chemicals, whether phosphate, chelate or the polymer, condition the calcium and magnesium within the feedwater. Chelates and polymers form the soluble complexes with the hardness, but the phosphates precipitate the hardness.

Sludge conditioners are also used to aid within the conditioning of precipitated hardness. These conditioners are selected in order that they are both effective and stable at boiler operating pressures. Synthetic organic materials are used as antifoam agents. For feedwater oxygen scavenging, the various chemicals used are sodium sulfite and hydrazine.

Condensate system protection may be accomplished by the use of volatile amines or the volatile filming inhibitors. A reputable company supplying treatment chemicals should be consulted. Those companies supply the chemical formulations under their brand names and they give details on the dosage and methods.

The Internal Treatment for Sulfates

The boiler temperature makes the calcium and magnesium sulfates within the feedwater insoluble. With the phosphates used as internal treatment, calcium reacts with the phosphate producing hydroxyapatite, that is much easier to condition than calcium sulfate.

With chelates or polymer used as internal treatment, the calcium and the magnesium react with these materials producing soluble complexes that are simply removed by blowdown.

The Internal Treatment for Silica

When silica is present in the feedwater, it tends to precipitate directly as scale at the hot spots on the boiler metal or combines with the calcium forming a hard calcium silicate scale. The boiler water alkalinity has to be kept high enough to carry the silica in solution in the internal treatment for silica.

Magnesium, present in most waters, it precipitates some of the silica as sludge. Special organic materials or synthetic polymers are used to condition the magnesium silicate from adhering to the boiler metal.

The Internal Treatment for Sludge Conditioning

The internal treatment for hardness leads to insoluble precipitates in the boiler that form sludge. Additionally, insoluble corrosion particulates (metal oxides) are transported to the boiler by condensate returns and from pre-boiler feedwater corrosion leading to suspended solids.

Suspended solids, carried to the boiler by feedwater or subsequently formed inside the boiler, adversely affect both boiler cleanliness and steam purity. This solid has a varying tendency to deposit on the boiler metal. Conditioners prevent these solids from depositing and forming corrosive or insulating boiler scale.

Some of the principal types of the sludge conditioners are:

- Lignins are effective on the phosphate type sludge.

- Tannins fairly effective on high hardness feedwater.

- Starches effective on high silica feedwater and where oil contamination is a problem.

- Synthetic polymers Highly effective sludge conditioners for all types of sludges.

Advantages

The internal treatment is simple basically and with the help of a qualified consultant an effective program is simply established.

The scales or deposits, corrosion and carryover are minimized thereby improving efficiency and reducing the energy consumption, then preventing tube failures and unscheduled costly repairs and reducing deposits, corrosion and contamination in the downstream equipments or processes.

5.3.1 Softening of Hard Water

The water used in boiler for steam production should be pure otherwise it would cause boiler troubles like scale, sludge, priming and foaming etc.

When the hard water sample is passed through the I-Cylinder (acidic resin) calcium and magnesium ions are replaced by hydrogen ions of the acidic resin.

$$RH_2 + Ca^{2+} \rightarrow RCa + 2H^+$$

Acidic resin,

$$RH_2 + Mg^{2+} \rightarrow RMg + 2H^+$$

Acidic resin,

When this water is passed through the II-Cylinder chloride, bicarbonate and sulphate ions are replaced by the hydroxide ions of the basic resins.

$$R'(OH)_2 + 2Cl \rightarrow R'Cl_2 + 2OH^-$$

$$R'(OH)_2 + 2HCO_3^- \rightarrow R'(HCO_3)_2 + 2OH^-$$

$$R'(OH)_2 + SO_4^{2-} \rightarrow R'SO_4 + 2OH^-$$

Thus all the ions responsible for hardness are removed from water. The H^+ and OH^- ions combine together to form water.

$$H^+ + OH^- \rightarrow H_2O$$

The quality of water obtained by this method is equivalent to distilled water.

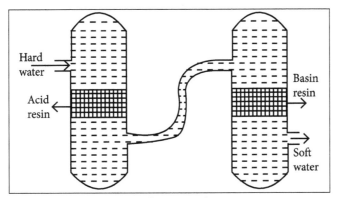

Cation and anion exchanger.

Regeneration of Acid Resin and Basic Resin

After a long use, the acidic resin can be regenerated by the addition of a strong solution of Hydrochloric acid.

$$RCa + 2HCI \rightarrow RH_2 + CaCl_2$$

The basic resin after a long use, can be regenerated by the addition of a strong solution of NaOH.

$$R'CI_2 + 2OH^- \rightarrow R'(OH)_2 + 2CI^-$$

$$R'(HCO_3)_2 + 2OH^- \rightarrow R'(OH)_2 + 2HCO_3^-$$

$$R'SO_4 + 2OH^- \rightarrow R'(OH)_2 + SO_4^{2-}$$

Advantages

- In this method, both types of hardness are removed.

- The quality of water obtained is equivalent to distilled water.

- There is no wastage of water.

Calgon Conditioning

Calgon is the trade name of sodium hexa meta phosphate – $Na_2\left[Na_4\left(PO_3\right)_6\right]$. With calcium ions it forms a soluble complex and prevents scale and sludge formation. It is used for high and low pressure boilers.

$$2CaSO_4 + Na_2\left[Na_4\left(PO_3\right)_6\right] \rightarrow Na_2\left[Ca_2\left(PO_3\right)_6\right] + 2Na_2SO_4$$

In calgon conditioning the added calgon forms soluble complex, compound with $CaCO_3$. There by it prevents the formation of scale and sludge in boiler.

Since the complex formed is soluble, it does not cause any issues in the boiler. On the other hand, in phosphate conditioning sodium phosphate is added to the boiler water in order that calcium phosphate precipitates if formed.

Although the precipitate is non-adherent and soft, yet it has to be removed by frequent blow down operation. Hence, calgon conditioning is definitely better than phosphate conditioning.

Lime-Soda Process

In this Method, the soluble calcium and Magnesium salts in water are chemically converted into insoluble Compounds by adding calculated amounts of lime $Ca(OH)_2$ and soda Na_2CO_3 calcium carbonate $CaCO_3$ and magnesium hydroxide $Mg(OH)_2$. So Precipitated, are filtered off.

Cold Lime-Soda Process

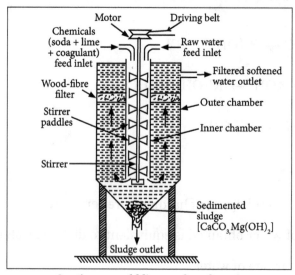

Continuous cold lime-soda softener.

In this method, calculated quantity of chemicals and water, along with accelerators and coagulators are added to a tank fitted with a stirrer. On vigorous stirring, thorough mixing takes place. After softening the soft water rises upwards and the heavy sludges settle down.

The softened water passes through a filtering media ensuring complete removal of the sludge and finally the filtered water flows out through the top.

Cold lime soda process is used for partial softening of municipal water, for softening of cooling water etc. In actual purpose, magnesium hardness is brought down to almost zero but calcium hardness remains about 40 ppm.

Method

Raw water and calculated quantities of chemicals (Lime + soda + coagulant) are fed from the top into the inner vertical circular chambers, fitted with a vertical rotating shaft carrying a number of paddles.

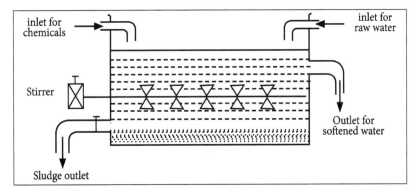

As the raw water and chemicals flow down there is a vigorous stirring and continuous mixing, whereby softening of water takes place. As the softened water comes into the outer chamber of the lime the softened water reaches up.

The softened water then passes through a filtering media to ensure complete removal of sludge. Filtered soft water finally flow out continuously through the outlet at the top sludge settling at the bottom of the outer chamber.

Hot Lime-Soda Process

This process is similar to the cold lime-soda process. Here the chemicals along with the water are heated near about the boiling point of water by exhaust steam. As the reaction takes place at high temperature, it has the following advantages.

Advantages

- The precipitation reaction becomes almost complete.

- The reaction takes place faster.

- The sludge settles rapidly.

- No coagulant is needed.

- Dissolved gases (which may cause corrosion) are removed.

- Viscosity of soft water is lower, hence filtered easily.

- Residual hardness is low compared to the cold process.

Hot Lime-soda Process Consists of Three Parts

- Reaction tank in which complete mixing of the ingredients takes place.

- Ionical sedimentation vessel where the sludge settles down.

- Sand filter where sludge is completely removed.

The soft water from this process is used for feeding the boilers.

Advantages

- Lime soda process is economical.

- The process improves the corrosion resistance of the water.

- Mineral content of the water is reduced.

- pH of the water rises, which reduces the content of pathogenic bacteria.

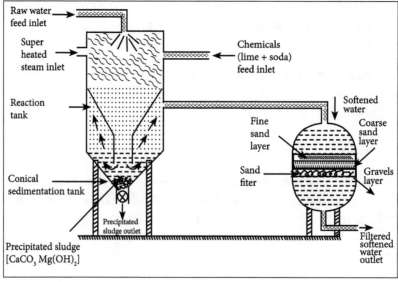

Continuous hot lime-soda softener.

Disadvantages

- Huge amount of sludge is formed and disposal is difficult.

- Due to residual hardness, water is not suitable for high pressure boiler.

Zeolite or Permutit Process

The name Zeolite is derived from Greek words (Zien + Lithos) which mean "boiling stone". The chemical structure of Sodium Zeolite may be represented as $Na_2O, Al_2O_3, xSiO_2, yH_2O$ where $(x = 2-10$ and $y = 2-6)$.

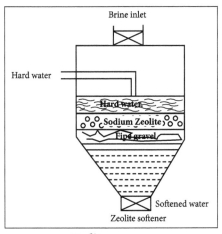

Zeolite process.

Zeolite or permutit is hardness producing Ca^{+2} and Mg^{2+} ions in water Zeolites are of two types:

- Natural Zeolites: Natural Zeolites are more durable. E.g., Non-porous green sands, natrolite, $Na_2O, Al_2O_3, SiO_2, 2H_2O$.

- Synthetic Zeolites: Synthetic Zeolites are porous and possess a gel structure. These Zeolites are prepared from solution of sodium silicate and aluminium hydroxide. Synthetic Zeolites have higher exchange capacity per unit weight.

Sodium Zeolites are used in water softening and are represented as Na_2Z (Z stands for the insoluble zeolite radical framework). These are also known as base exchangers.

Zeolites Process for Treating Water

Zeolites water softeners are made in both pressure type and gravity type. A Zeolite softener consists of steel tank packed with a thick layer of permutit. The water enters at the top and passes through the bed of Zeolite.

It operates on the principle involving alternate cycles of softening and regeneration.

The water is softened by passing it through the Zeolite bed, where Ca^{+2} and Mg^{2+} ions are removed from the water by Zeolite and simultaneously releasing equivalent amount of Na^+ ion exchange.

Process: In this process hard water passes at a specified rate through a bed of active granular sodium zeolite present in a zeolite as CaZ and MgZ respectively while the outgoing water contains equivalent amount of sodium salts. The chemical reactions taking place in zeolite softener are:

$$Ca\left(HCO_3\right)_2 + Na_2Z \rightarrow CaZ + 2NaHCO_3$$

$$Mg\left(HCO_3\right)_2 + Na_2Z \rightarrow MgZ + 2NaHCO_3$$

$$CaSO_4 + Na_2Z \rightarrow CaZ + Na_2SO_4$$

$$CaCl_2 + Na_2Z \rightarrow CaZ + 2NaCl$$

Small quantities of iron and manganese present as the divalent bicarbonates may also get removed simultaneously.

$$Fe\left(HCO_3\right)_2 + Na_2Z \rightarrow Fe_2 + 2NaHCO_3$$

$$Mn\left(HCO_3\right)_2 + Na_2Z \rightarrow Mn_2 + 2NaHCO_3$$

Regeneration

After some time, when the zeolite is completely changed into calcium and magnesium zeolites it gets exhausted (saturated with Ca^{2+} and Mg^{2+} ions) and it ceases to soften water. It can be regenerated and reused by treating it with a 10% brine (sodium chloride) solution.

$$CaZ + 2NaCl \rightarrow Na_2Z + CaCl_2$$

$$MgZ + 2NaCl \rightarrow Na_2Z + MgCl_2$$

Advantages of Zeolite Process

- It removes the hardness almost completely.

- The equipment used is compact occupying a small space.

- The process automatically adjusts itself for variation in hardness of incoming water.

- The process does not involve any type of precipitation thus no problem of sludge formation occurs.

- The plant can be installed in the water supply line itself decreasing the cost of pumping.

- It requires less time for softening.

Disadvantages of Zeolites Process

- The outgoing water (treated water) contains more sodium salts.

- The method only replaces Ca^{2+} and Mg^{2+} ions by Na^+ ions.

- Zeolite process leaves all the acidic ions (like HCO^{3-} and CO_2^{-3} as such in the softened water.

- High turbidity water cannot be softened efficiently by Zeolite process.

Limitations of Zeolite Process

- The water must be free from turbidity and suspended matter. Otherwise Zeolite bed (Permutits) will be clogged and the rate of flow will be decrease.

- Hot water should not be used as the Zeolite tend to dissolve in it.

- Colored ions such as Mn^{2+} and Fe^{2+} must be removed first because these ions produce manganese and ion Zeolites which cannot be easily regenerated.

Ion Exchange or Demineralization Process

Ion exchange or demineralization process removes almost all ions present in the hard water. The soft water is produced by lime-soda and zeolite processes does not contain hardness producing Ca^{2+} and Mg^{2+} ions but it will contain other ions like Na^+, K^+, SO_4^{2-}, Cl^- etc.

On the other hand demineralized (DM) water does not contain both anions and cations. Thus a soft water is not demineralized water whereas demineralized water is soft water.

This process is carried out by using ion exchange resins, that are long chain, cross linked, insoluble organic polymers with a micro process structure. Therefore the functional groups attached to the chains are responsible for the ion exchanging properties.

Applications

- Removal of non-metal inorganic.

- Water softening.

- Removal or recovery of metal.

1. Cation Exchanger

Resins containing acidic functional groups $\left(- COOH, - SO_3H\right)$ are capable of exchanging their H^+ ions with other cations of hard water. Cation exchange resin is represented as RH_2.

Examples: Sulphonated coals.

Sulphonated polystyrene.

$$R - SO_3H; R - COOH = RH_2$$

2. Anion Exchanger

Resins containing basic functional groups $(-NH_2, -OH)$ are capable of exchanging their anions with other anions of hard water. Anion exchange resin is represented as $R\left(OH\right)_2$.

Examples: Cross-linked quaternary ammonium salt.

Urea-formaldehyde resin.

$$R - NR_3OH; \ R - OH; \ R - NH_2 = R\left(OH\right)_2$$

Process

The hard water is first passed through a cation exchange which absorbs all the cations like Ca^{2+}, Mg^{2+}, Na^+, K^+, etc. present in the hard water.

$$RH_2 + CaCl_2 \rightarrow RCa + 2HCl$$

$$RH_2 + MgSO_4 \rightarrow RMg + H_2SO_4$$

$$RH + NaCl \rightarrow RNa + HCl$$

The cation free water is then passed through a anion exchange column which absorbs all the anions like Cl^-, SO_4^{2}, HCO_3^- etc., present in the water.

$$R'\left(OH\right)_2 + 2HCl \rightarrow R'Cl_2 + 2H_2O$$

$$R'\left(OH\right)_2 + H_2SO4 \rightarrow \ 'R'SO_4 + 2H_2O$$

The water coming out of the anion exchanger completely free from cations and anions. This water is known as demineralized water or deionized water.

Equipment

The ion exchange resin is contained in a vessel with a volume of several cubic feet. Retention components at the top and bottom consist of screens, slotted cylinders or alternative suitable devices with openings smaller than the resin beads to prevent the resin from escaping from the vessel.

When the resin bed may be a uniform mixture of cation and anion resins in a volume. This arrangement is known as a mixed-bed resin, as opposed to an arrangement of cation and anion resins in discrete layers or separate vessels.

The use of various volumes of the two kinds of resins is due to the difference in exchange capacity between cation and anion resins. The exchange capacity is the amount of impurity that a given amount of resin is capable of removing and it has units of moles, equivalents or moles.

The anion resin is less dense than the cation resin, so, it has a smaller exchange capacity and a larger volume is required for anion resins than for the cation resins to obtain equal total exchange capabilities.

Working

- All the cations and anions are completely removed by two column of cation exchange column and anion exchange column filled with resins.

- Resins are long chain, insoluble, cross linked, organic polymers. There are two types:

  ○ Cation exchange resins $- RH^+$.

    (e.g) Sulphonated coals, RSO_3H.

  ○ Anion exchange resin $- R'OH^-$.

    (e.g) Urea formaldehyde, Amines $R - NH_2$.

- The water is fed into cylinder-I where all the cations are replaced by RH_2 resins,

  $$2RH^+ + Ca^{2+} \rightarrow R_2Ca^{2+} + 2H^+.$$

- The cation free water is fed to cylinder II where all the anions are replaced,

  $$2ROH^- + SO_4^{2-} \rightarrow R_2SO_4^{2-} + 2\,OH^-.$$

- Therefore the resultant water is free from all types of ions.

Ion exchange process.

Advantages of Ion Exchange Method

- It can be used for high pressure boilers also.

- It can treat highly acidic or alkaline water.

- We can get pure water with hardness as low as 2 ppm.

Drawbacks of Ion Exchange Method

- Expensive.

- Fe, Mn cannot be removed as they form complexes with resins.

- It cannot be used for turbid water as they clog the resins.

Regeneration

When the cation exchange resin is exhausted, it can be regenerated by passing a solution of dil.HCL or dil.H_2SO_4.

$$RCa + 2HCl \rightarrow RH_2 + CaCl_2$$

$$RNa + HCl \rightarrow RH + NaCl$$

Similarly when the anion exchange resin is exhausted, it can be regenerated by passing a solution of dil.NaOH.

$$R'Cl_2 + 2\,NaOH \rightarrow R'(OH)_2 + 2\,NaCI$$

Advantages

Highly acidic or alkaline water can be treated by this method.

Disadvantages

- The equipment is costly.

- More explosive chemicals are needed for this process.

- Water containing turbidity, Fe and Mn cannot be treated because turbidity reduces the output and Fe, Mn form stable compound with the resins.

Numerical Problems

1. A pond water sample contains 100 mg of Ca(HCO) per liter, Let us calculate the hardness of water in terms of CaCO equivalent.

Solution:

Given:

100 mg of Ca(HCO) per liter

In the present case,

Mass of Ca(HCO) (w) = 100 mg/liter

$$\therefore \text{ Equivalent of CaCO} = \frac{w \times 50}{E} = \frac{120 \times 50}{81} = 74.07 \text{ mg / litre or ppm.}$$

Equivalent mass of Ca(HCO)(E) = 81

2. A sample of water contains 200 mg of Ca^+ per liter. Let us calculate the hardness of the sample in terms of CaCO equivalent.

Solution:

Given:

In the present case,

Mass of Ca^+ (w) = 200 mg/liter

Equivalent mass (E) of Ca^+ = 20

Equivalent of CaCO = w x 50/E = 200 x 50/20 = 500 mg/liter or ppm.

3. Hardness of 4,500 liters of water was removed completely by zeolite softener. This

zeolite required 30 liters of 100 gm/lit of NaCl to regenerate. Let us calculate the hardness of water.

Solution:

Using regeneration reaction,

$$CaZe + 2NaCl \rightarrow Na_2Ze + CaCl_2$$

i.e. $2NaCl \equiv CaCl_2 \equiv CaCO_3$

$$2(58.5) \text{ gm} \equiv 111 \text{ gm} \equiv 100 \text{ gm of } CaCO_3$$

Now,

Quantity of NaCl in regeneration,

$$= 30 \times 100$$

$$= 3000 \text{ gm NaCl}$$

Thus,

$$(2 \times 58.5) \text{ gm NaCl} \Rightarrow 100 \text{ gm } CaCO_3$$

$$\therefore 3000 \text{ gm NaCl} = \times \times = \times$$

$$= 2564.102 \text{ gm of } CaCO_3$$

$$= 2564102 \text{ mg of } CaCO_3$$

$\therefore$ 4500 liter of hard water= 2564102 mg $CaCO_3$

1 liter of hard water = mg of $CaCO_3$

$$= 56.98 \text{ mg of } CaCO_3 = 56.98 \text{ ppm}$$

$\therefore$ Hardness of water sample = 569.98 ppm.

4. 15,000 liter of hard water was passed through a zeolite softener. The exhausted zeolite required 120 liter of NaCl having 30 g / liter of NaCl. Let us calculate the hardness of water.

Solution:

Let the hardness of the water sample be x mg/l

Now,

1 liter of NaCl contains 30 g of NaCl = 30000 mg NaCl

∴ 120 liter of NaCl contains 120 x 30000 = 3600000 mg of NaCl

58.5 mg of NaCl = 50 mg of $CaCO_3$ equivalent hardness

= 3600000 ×

= 30,76,923 mg of $CaCO_3$ equivalents hardness

But the total quantity of the water sample = 15,000 liter

∴ 15000 liter of water = 30,76,923 mg of $CaCO_3$

∴ 1 liter of water = mg of $CaCO_3$

= 205 mg/l of $CaCO_3$

= 205 ppm $CaCO_3$

∴ Hardness of water sample = 205 ppm.

5. Let us calculate the quantities of lime and soda (90 % pure each) required for softening 25000 liter of hard water containing following ions/chemicals.

$HCO_3^- $ = 12.2 ppm; CO_2 = 4.4 ppm

Ca^{++} = 30 ppm; Mg^{++} = 21.6 ppm

Fe_2O_3 = 15.4 ppm; H_2SO_4 = 4.9 ppm

Solution:

Calculation of $CaCO_3$ equivalents for impurities

Salt/impurity	Qty in mg/lit	Multiplication factor	$CaCO_3$ equivalent ppm	Requirement of Lime (L) and / or Soda (S)
Ca^{2+}	30 ppm	30 ×	75	S
Mg^{2+}	21.6 ppm	21.6 ×	90	L + S
H_2SO_4	4.9 ppm	4.9 ×	5	L + S
CO_2	4.4 ppm	4.4 ×	10	L
-	12.2 ppm	12.2 ×	10	Add in L and subtract in S
Fe_2O_3	15.4 ppm	Does not contribute		

Calculation of Quantity of Lime, Required for Softening of Water

Quantity of Lime

$$L = [\text{Temporary Ca}^{+2} + 2 \times \text{Temporary Mg}^{+2} + \text{Permanent}$$
$$\left(\text{Mg}^{+2} + \text{Fe}^{+2} + \text{Al}^{+3} + \text{H}^+\left(\text{HCl or H}_2\text{SO}_4\right)\right) + \text{CO}_2 + \text{HCO}^-_3 - \text{NaAlO}_2\,] \times \times \text{ kg}$$

all in terms of their $CaCO_3$ equivalents.

$$= [90 + 5 + 10 + 10] \times \times \text{ kg}$$

$$= 2.3638 \text{ kg}$$

Calculation of Quantity of Soda Required for Softening Water

Quantity of Soda

$$S = \left[\text{Permanent}\left(\text{Ca}^{+2} + \text{Mg}^{+2} + \text{Al}^{+3} + \text{Fe}^{+2} + \text{H}^+\left(\text{HCl or H}_2\text{SO}_4\right)\right) - \text{HCO}^-_3\right] \times \times \text{ kg}$$

all in terms of their $CaCO_3$ equivalents.

$$= \left[\text{Perm of Ca}^{+2} + \text{Perm of Mg}^{+2} + \text{H}_2\text{SO}_4 - {}^-\right] \times \times \text{ kg}$$

$$= [75 + 90 + 5 - 10] \times$$

$$= 4.7111 \text{ kg}$$

Quantity of lime required = 2.3638 kg

Quantity of soda required = 4.7111 kg.

5.4 Water for Drinking Purposes, Purification, Sterilization and Disinfection

Water for Drinking Purpose (Potable Water)

The municipal water is mainly used for drinking purposes and for cleaning, washing and other domestic purposes. Therefore the water that is fit for drinking purposes is called potable water.

1. Characteristics of Potable water

- It should not have turbidity and other suspended Impurities.

- It should be colorless, odorless and tasteless.

- It should not contain toxic dissolved impurities.

- It should be free from the germs and bacteria.

- It should not be corrosive to the pipe lines.

- It should not stain the clothes.

- It should be moderately soft.

2. Standards of Drinking Water as Recommended by WHO

Parameter	WHO Standards
pH	6.5 -9.2
BOD	6
COD	10
Calcium	100ppm
Cadmium	0.01
Chromium	0.05
Ammonia	0.5ppm
copper	1.5ppm
Mercury	0.1ppm

3. Water Quality Standards in India

Parameter	Standards
pH	6.3 -9.2
Total Hardness	600 ppm
Chorides	1000 ppm
Cyanide	0.05 ppm
Cadmium	0.01 ppm
Chromium	0.05 ppm
Zinc	15 ppm
copper	1.5ppm
Mercury	0.001ppm
Nitrate	45 ppm
sulphate	400 ppm

Purification

Water purification is the process of removing undesirable chemicals, biological contaminants, suspended solids and gases from contaminated water. The goal is to produce water fit for a specific purpose.

Most water is disinfected for human consumption (drinking water), but water purification may also be designed for a variety of other purposes, including fulfilling the requirements of medical, pharmacological, chemical and industrial applications.

The methods used include physical processes such as filtration, sedimentation and distillation, biological processes such as slow sand filters or biologically active carbon, chemical processes such as flocculation and chlorination and the use of electromagnetic radiation such as ultraviolet light.

Purifying water may reduce the concentration of particulate matter including suspended particles, parasites, bacteria, algae, viruses, fungi, as well as reducing the amount of a range of dissolved and particulate material derived from the surfaces that come from runoff due to rain.

The standards for drinking water quality are typically set by governments or by international standards. These standards usually include minimum and maximum concentrations of contaminants, depending on the intended purpose of water use.

Sterilization and Disinfection

Sterilization Techniques

There are several ways to sterilize the water. The technique that can be applied for the sterilization depends on the situation i.e. the type of water available and the items we have both in the original water as well as the tools available. Having a knowledge about these methods can help in avoiding mistakes.

Boiling

Boiling water is the most commonly used method for cleaning up dirty water and it has a lot of benefits. All that is required is a pan and a lot of heat, which can kill many of the common waterborne bacteria and make it much closer to safe drinking water.

To be more effective, the water must actually be boiling in other words, a rolling boil with lots of big bubbles breaking the surface of the water very rapidly. Continue this for several minutes.

If boiling is our only option, it is certainly better than nothing. However, boiling does not remove all the possible contaminants from water and does nothing about the sediment or solid particles in the water.

Distillation

Distilling water takes a bit more set up, but is worth it in terms of cleaner water that is sediment free. In order to distill water, we can use a pan for boiling, a collection tray for the steam and another pan, pitcher or other receptacle for storing the treated water.

For example, by using a wire, attach a large piece of tin foil to two pans. The foil should be raised above the pans and sloping downward from the one over the heat source, to the one for collection. Put the water into the pan over the heat source.

As the water boils, stem will rise, condense on the tin foil and then run down into the second pan for collection. Distillation is not a perfect solution either, as some contaminants will still get through in this process.

Chlorine

When it comes to killing of bacteria, treating the water with chlorine bleach is probably the best option. Use 8 drops of bleach per gallon of water, stir and then allow the water to sit with the chlorine for at least one hour.

Of course, the downside to this treatment option is that we should have chlorine. Make sure that the chlorine bleach that we have for water treatment is the plain and unscented kind so that we don't end up drinking unwanted chemicals.

Chlorination

Chlorination is addition of chlorine. Chlorine is added to water in the pH range of 6.5 to 7. When chlorine is added to water, it forms HCL and HOCl. The hypochlorous acid enters into the living cells of bacteria and destroy them.

$$H_2O + Cl_2 \rightarrow HCl + HOCl$$

Hypochlorous acid

Other sterilizing agents used are chloramines, bleaching powder etc. The advantage of using chloramines is that it does not evaporate out easily and can be carried over to a longer distance along with the water.

Chlorination is the most common disinfection method in drinking water treatment where they can be done at any stage throughout the water treatment process. Each point of chlorine application shall control a different water contaminant concern, therefore provides a complete spectrum of treatment from the time water enters the treatment facility till the time it leaves.

The chlorination process is integrated into water treatment plants as a primary or secondary disinfection method.

Advantages

- Inexpensive.
- Effective in purifying water from pathogens and some inorganic compounds (iron, manganese, hydrogen sulphide).

- Non-toxic (in free chlorine form).

- Reduces taste and odor problems caused by algae and some chemical compounds.

Disadvantages

- CDBP's may be toxic.

- There are taste and odor problems with chlorine and CDBP.

Types of Chlorination

- Plain chlorination.

- Pre – chlorination.

- Post – chlorination.

- Double chlorination.

- Break point chlorination.

- Super chlorination.

- Dechlorination.

Plain Chlorination

It can treat water with turbidity in the range 20 to 30 mg/l. It is used in emergency period - Chlorine required $\geq$ 0.5 mg/l.

Pre Chlorination

Chlorine added to water before filtration (sedimentation) improves coagulation. It reduces odour, taste, algae and other microbes. It's dosage limit 5 to 10 mg/l.

Post Chlorination

It involves applying chlorine at end of process - residual chlorine is(0.1 to 0.2 mg/l) after the contact period of 20 minutes.

Double Chlorination

Combination of Pre and Post chlorination.

Break Point Chlorination

BPC is excess chlorine dose added beyond the limit where additional chlorine will appear as free residual chlorine.

Super Chlorination

Chlorine dosage beyond BPC. It is used when water is highly polluted - later excess residual chlorine is removed by De-chlorination.

De-chlorination

Used when super chlorination is practiced. It is done simply by aeration or using chemicals.

De-chlorination agents are as follows:

- Sulphur dioxide gas.

- Activated carbon.

- Sodium thio sulphates.

- Sodium metabisulphate.

- Ammonia as NH_4OH.

Iodine

We can also use iodine. Make sure to use actual iodine and not any of a multitude of over-the-counter antiseptics or mixtures that contain iodine.

Like using the chlorine bleach, simply add the iodine and wait for an hour to let it do its magic. we will have to use up to one tablet per quart of water.

Store Bought Filtration

Safe water is a big deal and as a result it is also a big business. There are a whole multitude of water filtration systems on the market that cleanse the water of a whole variety of things. They even have water filtration "straws" for backpackers and others who may need to filter water on the move.

Powdered Charcoal

If we are in the woods, something we are sure to have access to is fire. Not only can that fire provide warmth and cooking power, but the ashes can help to purify the water.

Grind up clean ashes from the fire. In other words, ashes that are from untreated wood that have just been burned and which are not sitting around for a long time. Place the ashes in a strainer, place the strainer atop a funnel and pour the water through.

While this method alone will probably not be sufficient, it serves as a great backup to boiling, distilling or bleaching water. Once have done one of those things first, the charcoal may get some of the contaminants in the water that the first method may have left behind.

Water is Life

Never take for granted the easy access to clean water that most of us in the western world enjoy is present everywhere. In other parts of the world they regularly deal with dysentery, illness and even death because of a lack of clean portable drinking water.

Make sure we are not dependent on any on-grid systems to keep this life-giving substance flowing to us and know how to make sure our family is getting the safest water possible no matter what the circumstances.

Disinfection

Water disinfection means the removal, deactivation or killing of pathogenic microorganisms. Microorganisms are destroyed or deactivated, resulting in termination of growth and reproduction. When microorganisms are not removed from drinking water, drinking water usage will cause people to fall ill.

Sterilization is a process related to disinfection. However, during the sterilization process all present microorganisms are killed, both harmful and harmless microorganisms.

Disinfection can be achieved by physical or chemical methods. Chemicals used in disinfection are called disinfectants.

Different disinfectants have different target ranges, not all disinfectants can kill all microorganisms. Some methods of disinfection such as filtration do not kill bacteria, they separate them out. Sterilization is an absolute condition while disinfection is not. The two are not synonymous.

A Disinfectant Should

- Not cause the water to become toxic or unpalatable.

- Be able to destroy all types of pathogens and in whatever number present in the water.

- Be safe and easy to handle.

- Destroy the pathogens within the time available for disinfection.

- Be easy to determine its concentration in the water.

- Function properly regardless of any fluctuations in the composition or condition of the water.

- Provide residual protection against recontamination.

- Function within the temperature range of the water.

For chemical disinfection of water the following disinfectants can be used:

- Chlorine (Cl_2).

- Chlorine dioxide (ClO_2).

- Hypo chlorite (OCl^-).

- Ozone (O_3).

- Halogens: bromine (Br_2), iodene (I).

- Bromine chloride (BrCl).

- Metals: copper (Cu^{2+}), silver (Ag_+).

- Kaliumpermanganate $(KMnO_4)$.

- Fenols.

- Alcohols.

- Soaps and detergents.

- Kwartair ammonium salts.

- Hydrogen peroxide.

- Several acids and bases.

For physical disinfection of water the following disinfectants can be used:

- Ultraviolet light (UV).

- Electronic radiation.

- Gamma rays.

- Sounds.

- Heat.

Disinfection commonly takes place because of cell wall corrosion in the cells of the microorganisms or changes in cell permeability, protoplasm or enzyme activity. These disturbances in the cell activity will not allow the microorganisms to multiply. This will also cause the microorganisms to die out. Oxidizing disinfectants also demolish organic matter in the water, causing a lack of nutrients.

Theory of Disinfection

Factors affecting disinfection.

Disinfection - chlorine dioxide, chloramines, ozonation, UV radiation.

Primary Disinfection

Primary disinfection applies a disinfectant in the drinking water treatment plant. The amount of chlorine needed and time needed to react and disinfect is called the Contact Time (CT) and is a product of the concentration of residual chlorine (mg/l) and the disinfectant contact time.

CT values required to achieve the necessary disinfection depends on the microorganism targeted, pH and temperature. Other design factors influencing the amount of chlorine required are the contact chamber design, adequate mixing and the presence of sunlight.

Secondary Disinfection

Secondary disinfection may be applied to the treated water as it leaves the treatment plant or at re-chlorination points throughout the distribution system, to introduce and maintain a chlorine residual in the drinking water distribution system. Overall, a chlorine residual provides two main benefits:

- It can limit the growth of bio-film within the distribution system and its associated taste and odour problems.

- A rapid drop in disinfectant residual may provide an immediate indication of treatment process malfunction or a break in the integrity of the distribution system.

Reverse Osmosis

- Removal of common salt (NaCl) from water is called desalination.

- Brackish water: Water containing dissolved salts with a peculiar salty taste.

- Osmosis: When two different concentrated solutions are separated by a semi permeable membrane due to osmotic pressure low concentrated solvent flows to higher one. This is known as osmosis.

- But when we apply an excess and opposite Hydro static pressure ($15\text{-}40\,\text{kg/cm}^2$) to overcome the osmotic pressure then higher concentrated solvent will flow to the lower one. This is known as reverse osmosis.

- During RO process, only the water flows across the membrane and it prevents the salt migration. So this method is also called as super filtration.

- The membrane is made up of cellulose acetate, cellulose butyrate, polymethacrylate.

- The driving forces in this phenomenon are called osmotic pressure. If a hydrostatic pressure in excess of osmotic pressure is applied on the higher concentration side the solvent flow reverses i.e., solvent is forced to move from higher concentration to lower concentration. This is the principle of reverse osmosis. Thus in reverse osmosis method pure water is separated from its dissolved solids.

- Using this method pure water is separated from sea water. This process is also known as super-titration. The membranes used are cellulose acetate, cellulose butyrate, etc.

Principle of Reverse Osmosis

Reverse Osmosis makes utilization of outside and additional vitality in light of the fact that the regular vitality of the Osmosis technique is inadequate. The higher the weight applied while utilizing the pump shall expand the weight on salt side and powers the water to experience the semi-porous film.

On that occasion when this happens, the water abandons with regards to 95% percent of dissolved salts at the base, going into the reject stream. The weight is normally high however everything depends upon the salt grouping of the water, whenever the focus in the food water is all the additional, then we have to apply more weight.

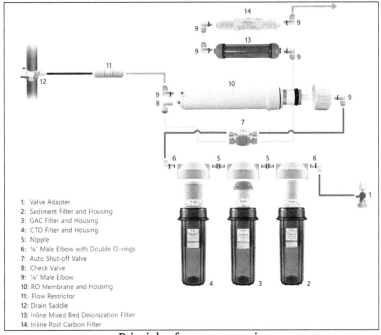

1: Valve Adapter
2: Sediment Filter and Housing
3: GAC Filter and Housing
4: CTO Filter and Housing
5: Nipple
6: ¼" Male Elbow with Double O-rings
7: Auto Shut-off Valve
8: Check Valve
9: ¼" Male Elbow
10: RO Membrane and Housing
11: Flow Restrictor
12: Drain Saddle
13: Inline Mixed Bed Deionization Filter
14: Inline Post Carbon Filter

Principle of reverse osmosis.

Reverse Osmosis Membrane Cleaning

The films during this system assume a key part and we should include in intermittent cleaning to build its life span. The cleaning should be done not less than 1 to 4 times in a year. Obviously, the number relies on upon the nature of food water.

In situations wherever the weight drops and salt section builds, then it gets to be key to clean the layer. We can clean it while it is joined or evacuate it and tidy it up site. Off-site film cleaning is considered to be more viable and yields great results.

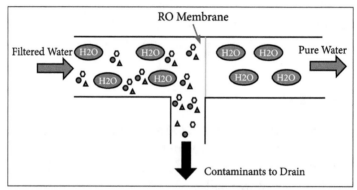

Reverse Osmosis Membrane Cleaning.

Reverse Osmosis Performance & Design Calculations

An RO system has instrumentation that displays quality, flow, pressure and typically other data like temperature or hours of operation. There are a handful of calculations which are used to decide the performance of an RO system and also for design considerations. In order to accurately measure the performance of an RO system the following operation parameters are required at a minimum:

- Permeate pressure.

- Feed pressure.

- Feed conductivity.

- Concentrate pressure.

- Permeate conductivity.

- Permeate flow.

- Feed flow.

- Temperature.

Salt Rejection %

$$\text{Salt Rejection \%} = \frac{\text{Conductivity of Feed Water} - \text{Conductivity of Permeate Water}}{\text{Conductivity of Feed}} \times 100$$

The higher the salt rejection, the better the system is performing. Therefore a low salt rejection will mean that the membranes need cleaning or replacement.

Salt Passage %

It is simply the inverse of salt rejection. This can be the amount of salts expressed as a percentage that are passing through the RO system. Thus, the lower the salt passage, the better the system is performing. A high salt passage will mean that the membranes require cleaning or replacement.

Salt passage % = (1 - Salt Rejection %)

Recovery %

Percent Recovery is that the amount of water which is being recovered as good permeate water.

$$\% \text{ Recovery} = \frac{\text{Permate Flow Rate (gpm)}}{\text{Feed Flow Rate (gpm)}} \times 100$$

Concentration Factor

The concentration factor is related to the RO system recovery and it is an important equation for RO system design.

Concentration factor = (1/(1-Recovery %))

As the degree of concentration increases, the solubility limits could also be exceeded and precipitate on the surface of the equipment as scale.

Flux

$$\text{Gfd} = \frac{\text{gpm of permeate} \times 1,440 \text{ min/day}}{\text{\# of RO elements in system} \times \text{square footage of each RO element}}$$

Mass Balance

The mass balance equation is used to help determine if our flow and quality instrumentation is reading properly or needs calibration. If our instrumentation is not reading correctly, then the performance data that we are collecting is useless.

The mass balance equation is:

Feed flow x Feed Conductivity = Permeate Flow x Permeate Conductivity + Concentrate Flow

* Concentrate Conductivity

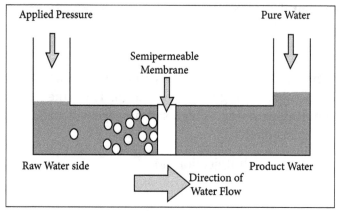

Reverse osmosis.

Advantages of Reverse Osmosis

- High life time.

- Removes ionic, non-ionic and colloidal silica impurities which cannot be removed by demineralization method.

- Low capital cost.

- The membrane can be replaced within a few minutes thereby providing uninterrupted water supply.

- Simple operational procedure.

Electrodialysis

Electrodialysis is based on the fact that the ions present in saline water migrate towards oppositely charged electrodes under the influence of an applied emf. The movement of ions takes place through ion-selective membranes. An electrodialysis cell consists of alternate cation- and anion-permeable membranes.

The cathode is placed near the cation-permeable membrane (C) and the anode is placed near the anion-permeable membrane (A). Under the influence of an emf applied across the electrodes, the anions (Cr) move towards the anode and the cations (NC) move towards the cathode.

The net result is the depletion of ions in the even-numbered compartments and concentration of ions in the odd-numbered compartments. Now the compartments with even-number are filled with pure water and the compartments with odd-number are with concentrated brine solution. Thus the salinity is removed from salt water.

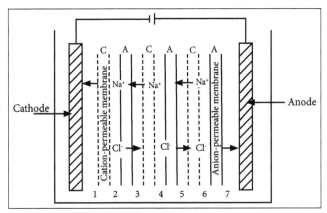

Electrodialysis

Electrodialysis is used to transport salt from one solution, the diluate, to another solution (concentrate) by applying an electric current. This is done in an electrodialysis cell providing all necessary elements for this process. The concentrate and diluate are separated by using the membranes, as shown in the figure below. An electric current is applied, moving the salt over the membranes.

Applications

- Desalination of salt water.

- Stabilization of wine.

- Whey demineralization.

- Pharmaceutical application.

- Pickling bath recycling.

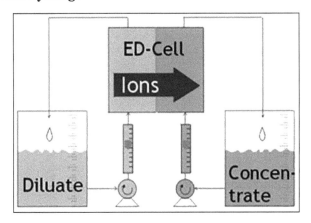

Inside an electrodialysis unit, the solutions are separated by using alternately arranged anion exchange membranes, permeable only for anions and cation exchange membranes, permeable only for cations.

By this, the two kinds of compartments are formed, distinguishing in the membrane type facing the cathode's direction. By applying a current, cations within the diluate move toward the cathode passing the cation exchange membrane facing this side and anions move towards the anode passing the anion exchange membrane.

A further transport of these ions, now being in a chamber of the concentrate, is stopped by the respective next membrane:

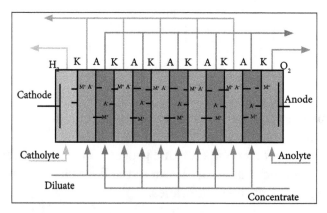

Electrodialysis Applications

Reduce Electrolyte Content

- Food products – soy sauce, milk, whey, fruit juice.

- Boiler feed water.

- Effluent streams.

- Electroless plating baths.

- Potable from brackish water.

- Sugar and molasses.

- Fiber reactive dyes.

- Nitrate from drinking water.

- Cooling tower water.

- Rinse water for electronics processing.

- Blood plasma to recover proteins.

- Pickle brines to recover flavor.

- Amino acids.

- Potassium tartrate from wine.

- Chloride purge in Kraft paper process.

- Photographic developer regeneration.

Recover Electrolytes

- Ag(I) salts from photographic waste.

- Pure NaCl from seawater.

- Zn(II) from galvanizing rinse water.

- Ni(II) from electroplating rinse waters.

- Amino acids from protein hydrolysates.

- Salts of organic acids from fermentation broth.

- HCl from cellulose hydrolysate.

- Acids from metal pickling baths and rinse.

Miscellaneous

- Metathesis.

- Salt splitting.

- Ion substitution.

- Concentrate reverse osmosis brines.

6

Chemistry of Engineering Materials and Fuel Cells

6.1   Refractories

Refractories are material that can withstand very high temperatures 3000°C or more without degrading or softening. Refractory materials include certain ceramics and super alloys and are used in heat insulation of furnaces, jet and rocket engines and parts of space vehicles such as the shuttle.

Characteristics

- Refractoriness: It is the ability to withstand very high temperature without deformation during operation.

- Strength or Refractoriness under load (RUL): They must possess high mechanical strength even at very high temperature and bear maximum possible load without breakage.

- Thermal expansion: A good refractory should have low thermal expansion under normal conditions.

- Thermal conductivity: In general, a good refractory must have low thermal conductivity to reduce heat losses by radiation. But when heat is to be supplied from outside the refractory must possess good conductivity.

- Porosity: A good refractory should have low porosity. In porous refractory, the molten metal and slag enters and weaken the structure. But porosity helps in thermal shock-resistance of refractories.

- Thermal spalling: It is breaking, cracking, peeling off or fracturing of the refractory under high temperature. A good refractory must have low thermal spalling.

- Chemical composition: A good refractory must be chemically inert with charge and slag.

Classification

Refractories are substance which can withstand high temperature without softening or deformation in shape. They are classified into three types:

- Acidic refractories. Ex: SiO_2, Al_2O_3.

- Basic refractories. Ex: CaO, MgO.

- Neutral refractories. Ex: Carbon.

The properties of refractories are as follows:

- Porosity.

- Fusion point.

- Spalling.

- Resistance to temperature change.

- Heat capacity.

- Thermal conductivity.

Refractoriness

Refractoriness is the ability to withstand high temperature without softening or deformation under particular service condition. Since most of the refractories are the mixtures of several metallic oxides, they do not have a sharp melting point.

So the refractoriness of a refractory is softening temperature and is expressed in terms of pyrometric cone equivalent(PCE). PCE is the number which represents the softening refractory specimen of standard dimension (38mm height and 19mm) composition.

Objective of PCE Test

- To determine the softening temperature of a test refractory material.

- To classify the refractories.

- To determine the purity of the refractories.

- To check whether the refractory can be used at particular servicing temperature.

Refractoriness is determined by comparing the softening temperature of a test cone with that of a series of segar cones. Segar cones are pyramid shaped standard refractory of definite composition and dimensions and hence it has a definite softening temperature.

A test cone is prepared from a refractory for which the softening temperature is to be determined, as the same dimensions of segar cones.

Then the test cone is placed in electric furnace. The furnace is heated at a standard rate of 10°C per minute, during which softening of segar cones occur along with test cone. The temperature at which the apex of the cone touches the base is taken as its softening temperature.

RUL - Refractoriness Under Load

The temperature at which a standard dimensioned specimen of a refractory undergoes 10% deformation with a constant load of 3.5 or 1.75 Kg/cm². The load bearing capacity of a refractory can be measured by RUL test. A good refractory should have high RUL value.

Dimensional Stability

The size and shape of the refractories is an important feature in design since it affects the stability of any structure. Dimensional accuracy and size is extremely important to enable proper fitting of the refractory shape and to minimize the thickness and joints in construction.

Thermal Spalling

It is breaking, cracking, peeling off or fracturing of the refractory under high temperature. A good refractory must have low thermal spalling.

Thermal Expansion

Bricks that have the coarse texture and low thermal expansion are more resistant to rapid thermal change. Less strain is developed in these kinds of bricks. When the bricks are being used for long time, they can melt easily.

Porosity

It is directly proportional to the resistance to chemical attack. The lower the porosity of the brick, the more difficult it is penetrated by molten fluxes and gases. Therefore the bricks having the lowest porosity have greatest strength, thermal conductivity and heat capacity.

Failures of Refractory Materials

We can easily summarize conditions, which lead to failures of a refractory materials as follows:

- Using a refractory material which does not have required heat, corrosion and abrasion resistance.

- Using refractory material of higher thermal expansion.

- Using a refractory of refractoriness less than that of the operating temperature.

- Using lower quality refractory bricks than the actual load of raw materials in products.

- Using basic refractory in a furnace in which acidic reactants and/or products are being processed and vice versa.

- Using refractories that undergo considerable volume changes during their use at high temperatures.

6.2   Lubricants

Whenever a machine works, its moving, sliding or rolling parts rub against each other resulting in the development of friction. This friction causes a lot of wear and tear of the concerned surfaces and hence such parts of machine require replacement at frequent intervals.

Furthermore, due to friction large amount of energy are dissipated in the form of heat and thus causes loss in the efficiency of machine. And also, the heat produced due to friction heats up and thus causes damage to the moving part.

A substance which is capable of reducing the friction between two surfaces which are sliding over one another is known as lubricant. The friction developed by the motion of two contacting surfaces is reduced by lubricant. In other words the loss of energy due to friction is considerably reduced by lubricant.

The lubricant acts in a number of manners:

- It forms a thin film between the rubbing surfaces. Thus, rubbing surfaces do not come in direct contact with each other.

- It acts as a coolant, as heat of friction is generated due to rubbing of surfaces.

- It avoids power loss in the internal combustion engines. Since, it acts as a seal by sealing the piston and cylinder wall at the compression rings hence, there is no leakage of gases even at high pressure in combustion chamber.

Functions

Important Functions of a lubricant include:

- It reduces loss of energy in the form of heat by acting as a coolant.

- It reduces wear and tear and surface deformation, by avoiding direct contact between the rubbing surfaces.

- It reduces the expansion of metal by local frictional heat.

- It reduces the maintenance as well as running cost of the machine.

- It minimizes the liberation of frictional heat and avoids seizures of moving surfaces.

- It minimizes rough relative motion of the moving (or sliding) parts.

- It increases the efficiency of the machine by reducing the waste of energy.

- It acts as a seal in internal combustion engines.

Theory and Mechanism of Lubrication

There are mainly three types of mechanism by which lubrication is done:

1. Thick Film or Fluid Film or Hydrodynamic Lubrication

In this mechanism, moving or sliding surfaces are separated by thick film of lubricant fluid, hence it is known as thick film or fluid film lubrication. This thick film of lubricant covers entire moving surfaces and fills irregularities.

Therefore, there is no direct contact between the surfaces of machine and consequently it reduces the wear. This is shown in Figure (a). Here only internal resistance is observed between the particles of lubricant, hence, chosen lubricant should have minimum viscosity under the working conditions.

Hydrodynamic friction occurs in the case of shaft running places like journal bearings. which is shown in Figure (b). Thick film lubrication hydrocarbon oils are considered as satisfactory lubricants. Hydrocarbon lubricants are blended with selected long-chain polymers to maintain viscosity of the oil throughout the year.

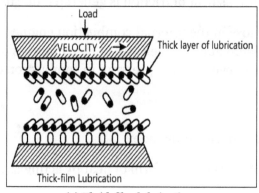

(a) Fluid-film lubrication.

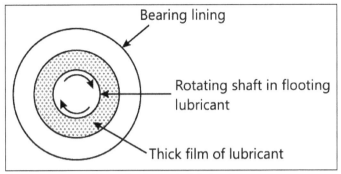

(b) Hydrodynamic lubrications.

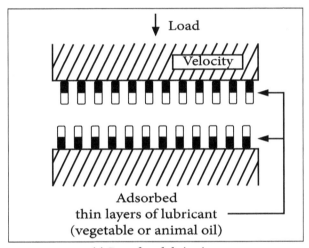

(c) Boundary lubrication.

2. Boundary or thin Film Lubrication

In this kind of lubrication moving surfaces are separated by thin layer of lubricant, which is absorbed by physical or chemical forces on the metallic surfaces as shown in Figure (c). Here continuous film of lubricant cannot persist due to the following reasons:

- A shaft starts moving from rest.
- The speed is very low.
- The load is very high.
- Viscosity of the oil is too low.

Vegetable oils, animal oils and their soaps possess the property of adsorption either physically or chemically to the metal surfaces and form a thin film of metallic soap, which acts as a good lubricant.

Fatty oils possess a greater adhesion property than mineral oil and to improve the oiliness of mineral oils small amount of fatty oils are added. Graphite and molybdenum disulphide are also used for boundary lubrication.

3. Extreme Pressure Lubrication

In this mechanism the moving or sliding surfaces are under very high pressure and speed hence this is known as extreme pressure lubrication. Under such conditions, a high local temperature is attained and due to this liquid lubricants fail to stick and may decompose or vaporize.

Special additives are added to mineral oils to meet the extreme pressure conditions and are called extreme pressure additives. Organic compounds which are having active radicals or groups such as chlorine, sulphur or phosphorous act as good additives.

These compounds react with metallic surfaces to form metallic chlorides, sulphides or phosphides as more durable films, capable of withstanding very high loads and temperatures.

Classification of Lubricants

On the basis of physical state, lubricants can be classified into three categories as listed hereunder.

1. Liquid Lubricants or Lubricating Oils

Lubricating oil also acts as a cooling medium scaling agent, corrosion preventer etc. along with reducing friction and wear.

According to origin, lubricating oils are classified into:

- Animal and vegetable oils.

- Mineral or petroleum oils.

- Blended oils.

i. Animal and Vegetable Oils

Vegetable and animal oils possess good oiliness but they are costly, they undergo oxidation easily, forming gummy and acidic products, get thickened on coming in contact with air, etc. Hence, they are rarely used as lubricant, but they are used as blending agent.

ii. Mineral or Petroleum Oils

Mainly mineral oils are obtained by distillation of petroleum. These are widely used lubricants because they are cheap, abundantly available and quiet stable under service conditions.

The hydrocarbon oil chain length varies between 12 to 50 carbon atoms. The shorter chain hydrocarbons have lower viscosity than longer chain hydrocarbons. When compared to animal and vegetable oils mineral oils possess poor oiliness, therefore, to increase oiliness high molecular weight compounds like oleic, steric acids are added.

iii. Blended Oils

In many modern machinery, no single oil serves as the most satisfactory lubricant. Improving of important properties by incorporating specific additives is known as blending of oils and blended oils give desired lubricating properties.

Properties of Good Lubricating Oil

A good lubricating oil must possess:

- Low pressure.

- High boiling point.

- Adequate viscosity to particular service conditions.

- Low freezing point.

- High oxidation resistance.

- Heat stability.

- Noncorrosive property.

- Stability to decomposition at the operating temperatures.

2. Greases or Semi-solid Lubricants

Lubricating grease is semi-solid, consisting of a soap dispersed throughout a liquid lubricating oil. The liquid lubricant may be a petroleum oil or even a synthetic oil and it may contain any of the additives for specific requirements.

Greases are prepared by saponification of fat with alkali, followed by adding hot lubricating oil while under agitation. The total amount of mineral oil added determines the consistency of the finished grease.

The structure of lubricating greases is like that of a gel. Soaps are gelling agents, which give an interconnected structure (held together by intermolecular forces) containing the added oil. The soap dissolves in the oil at high temperature, hence, inorganic solid, thickening agents are added to improve the heat resistance of grease. Greases have higher shear or frictional resistance than oils and can support much heavier loads at lower speeds.

Greases are used

- In situations where oil cannot remain in place, due to high load, low speed, intermittent operation, sudden jerks, etc., for example, tail axle boxes.

- In bearing and gears that work at high temperatures.

- In situations where bearing needs to be sealed against entry of dust, dirt, grit or moisture because greases are less liable to contamination by these.

- In situations where dripping or spurting of oil is undesirable, because unlike oils, greases if used do not splash or drip over articles being prepared by the machine, for example, in machines preparing paper, textiles, edible ankles. etc.

The main function of soap is that it acts as a thickening agent so that grease sticks firmly to the metal surfaces. However, the nature of the soap decides the temperature up to which the grease can be used, its consistency and its water and oxidation resistance. So greases are classified after soap is used in their manufacture.

3. Solid Lubricants

Graphite and molybdenum disulphide are the important solid lubricants. These are used in the following conditions.

- The operating temperature or load is too high.

- Blended lubricating oil or mixed grease is unacceptable.

- To avoid combustible lubricants.

Layered structure of graphite and sandwich-like structure of molybdenum disulphide are shown in Figures (a) and (b).

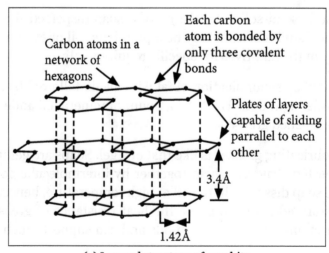

(a) Layered structure of graphite

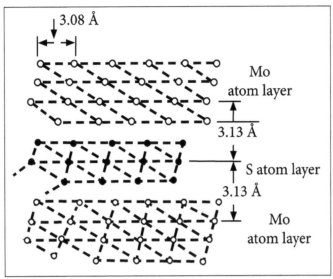

(b) Sandwitch-like structure of molybdenum disulphide.

Hence, the force to shear the crystals parallel to the layers is low and consequently the parallel layers slide over one another easily. Usually some organic substances are mixed with solid lubricants so that they may stick firmly to the metal surface.

Solid lubricants are used either in the dry powder form or mixed with water or oil. Graphite is the most widely used lubricant because it is very soapy to touch, non-inflammable and not oxidized in air below 375°C.

Graphite is used in the form of powder or suspension in oil or water with the help of emulsifying agent tannin. Graphite is dispersed in oil is called 'oildag' and when dispersed in water it is called 'aquadag'.

In the absence of air, it can be used up to very higher temperature. Graphite is used either in powdered form or as suspension. Graphite greases are used at higher temperature.

Properties of Lubricants

Viscosity

Viscosity is the most important property determining the behaviour of lubricants. It is the property of a fluid which resists its flow or shear. It may be defined as the property of a liquid or fluid by virtue of which it offers resistance to its own flow.

A liquid in a state of steady flow on a surface may be supposed to consist of a series of parallel layers moving one above the other. Any two layers will move with different velocities, top layer moves faster than the next lower layer, due to viscous drag (i.e., internal friction).

Consider two layers of a liquid separated by a distance, d and moving with a relative velocity difference, v, then, force per unit area (F) required to maintain this velocity difference is given by:

$$F = \eta = \frac{\upsilon}{d}$$

Where, η (eta) is constant of the liquid, called coefficient of viscosity. If v = 1 unit (e.g., cm/s), d = 1 unit (e.g., cm), then F=η. So, coefficient of viscosity may be defined as, the force per unit area required to maintain a unit velocity gradient between two parallel layers. The unit of viscosity is poise (dynes cm^{-2} or $g\ cm^{-1}s^{-1}$). Centipoise is however the unit in practice.

Cloud Point and Pour Point

When an oil is cooled slowly, the temperature at which it becomes cloudy or hazy in appearance, is called its cloud point, while the temperature at which the oil ceases to flow or pour, is called its pour point.

Most of the petroleum-based lubricants contain dissolved paraffin wax and asphaltic impurities. When the oil is cooled, these impurities will start crystallizing out giving a cloudy appearance.

Further, cooling of the oil causes the oil to crystallize out and minute crystals of wax get interlocked prevents the flow of the oil. This is the reason for considering the pour point. Thus, a knowledge of the cloud point will be helpful in deciding the lowest temperature up to which it can be used without any clogging.

The pour point helps to establish the lowest temperature at which the flow of the oil under gravity is reliable. Both these parameters may be useful in identifying the source of the oil. The pour point, however, is more important in the sense that it indicates the approximate minimum temperature at which the oil can be transferred by pouring or below which lubrication by gravity flow is unreliable.

The cloud and pour points can be used to assess the suitability of lubricants under cold conditions. The lubricants intended for use in machines at low temperatures should have low pour point. The presence of waxes in lubricating oils will tend to raise their pour points.

A simple way of determining the pour point is to read the maximum temperature at which the oil does not move when kept in a horizontal jar for five seconds. The pour point will be 5°F more than this temperature.

Flash Point and Fire Point

These two parameters were originally devised to assess the volatility of an oil which affects the consumption of an oil in an IC engine. The flash point is the lowest temperature at which sufficient vapours are given off by an oil to cause a momentary flash as the vapours come in contact with a test flame.

Flash point of burning oil (kerosene) is 73°F while that of power kerosene is 90°F. The fire point is the lowest temperature at which sufficient vapours are given off by an oil so that it bunts continuously for at least five seconds on being lighted by a test flame.

The flash point gives a rough indication of the tendency of the oil to volatilize and therefore is useful in process control for maintaining the uniformity of a product. In general, the fire points are 5 to 40° higher than the flash points.

A good lubricant should have its flash point at least above the temperature it is to be used. This provides protection against fire hazards during use, transport and storage of the lubricant. High flash and fire points are good characteristics of a lubricant.

A low flash point is indicative of evaporation losses and vapour locks in fuel passages of spark ignition engines. Oils of paraffinic base possess higher flash points as compared to naphthenic bases. Thus, the types of petroleum crude can be distinguished on the basis of flash points.

The flash and fire points do not have any bearing with the lubricating property of the oil, but they are important when oil is exposed to high temperature service. The flash and fire points are usually determined by using Pensky-Mannt's apparatus.

Oiliness

Oiliness of a lubricant is a measure of its capacity to stick on the surfaces of machine parts, under conditions of heavy pressure or load. When a lubricating oil of poor oiliness is subjected to high pressure, it has a tendency to be squeezed out of the lubricated machine parts, thereby its lubrication action stops.

On the other hand, lubricants, which have good oiliness stay in-between the lubricated surfaces, when they are subjected to high pressure. Oiliness is very important property of lubricants, particularly for extreme-pressure lubrication.

Mineral oils have got very poor oiliness, while vegetable oils have good oiliness. So, in order to improve the oiliness of mineral oils, additives like vegetable oils and higher fatty acids (such as oleic and stearic acids) are added to them. No direct test is available for measuring oiliness.

Decomposition and Corrosion Stability

Lubricating oils must be stable to decomposition at the operating temperatures. Oils are, usually, broken up by three chemical actions:

- Oxidation: It is the main destructive influence encountered. It is usually, partially prevented by adding certain antioxidants, which act as inhibitors.

- Hydrolysis: The presence of moisture in the oil causes hydrolysis of such components as esters in the lubricants, releasing alcohols and destructive fatty adds.

- Pyrolysis: It is the cracking of petroleum chains, due to the high temperature found within the engine bearings. Such a reaction is the main cause for the deposition of gummy and carbon sediments within the lubricant.

The harmful effects of contamination of the lubricant by these decomposition processes can be minimized by the use of an efficient system of oil filtration and periodic change of the oil. Corrosion stability of a lubricating oil is estimated by carrying out corrosion test.

A polished copper strip is placed in the lubricating oil for a specified time at a particular temperature. After the stipulated time, the strip is taken out and examined for corrosion effects. If the copper strip has tarnished, it shows that oil contains any chemically active substance.

A good lubricant should not affect the copper strip. To retard corrosion effects of oil, certain inhibitors are added to them. Commonly used inhibitors are organic compounds containing phosphorus, arsenic, antimony, chromium, bismuth or lead.

Emulsification

The ability of an oil to get intimately mixed with water forming an emulsion is called emulsification. Certain oils form emulsions with water easily. Emulsions have a tendency to collect din, grit, foreign matter, etc., thereby causing abrasion and wearing out of the lubricated parts of the machinery.

So, a good lubricating oil should form an emulsion with water, which breaks off quickly. The tendency of lubricant-water emulsion to break is determined by ASTM test. In this, 20 ml of oil is taken in a test-tube and steam at 100°C is bubbled through it, till the temperature is raised to 90°C.

The tube is then placed in a bath maintained at 90°C and the time in seconds is noted, when the oil and water separate out in distinct layers. The time in second in which oil and water emulsion separates out in distinct layers, is called steam emulsion number (SEN) A good lubricant should possess a low steam emulsion number.

Aniline Point

The tendency of a lubricant to mix with aniline is a measure of its aromatic content. This can be expressed in terms of aniline point which can be defined as the lowest temperature at which the oil is completely miscible with an equal volume of freshly distilled aniline.

The aniline point is inversely related to the aromatic content of the oil. Aromatic hydrocarbons have a tendency to dissolve natural rubber and certain types of synthetic rubbers. Low aromatic content is therefore desirable in the lubricants. A high aniline point suggests a higher percentage of paraffinic hydrocarbons.

The aniline point can be determined by mixing mechanically equal volumes of an oil sample and aniline in a tube. It is then heated until a homogeneous solution is formed. The tube is then allowed to cool at a controlled rate. The temperature at which the two phases separate distinctively is taken as the aniline point.

Volatility and Carbon Residue

Volatility

When a lubricating oil is used in heavy machinery working at high temperature, a portion of oil may vaporize, leaving behind a residual oil, which have different lubricating properties (like increased viscosity). Good lubricant should have low volatility.

The volatility of an oil is determined by an apparatus, called as vaporimeter, which consists essentially of a furnace heated by some fuel gas. In the centre of the furnace passes a coiled-form of copper tube, through which air can be passed.

A known weight of oil under-examination is taken in a platinum tray, which is then introduced into the copper tube. Dry air at a rate of 2 liter/minute is passed through the copper tube. After 1 hour of heating, the tray is taken out, cooled and weighed. The loss in weight is calculated as percentage of the original weight of oil taken.

Carbon Residue

Lubricating oils contain high percentage of carbon in combined form. On heating, they decompose depositing a certain amount of carbon. The deposition of such carbon in machine is intolerable, particularly in internal combustion engines and air-compressors.

A good lubricant should deposit least amount of the carbon in use. The estimation of carbon residue is, generally, carried out by Conradson method. A weighed quantity of oil is taken in a silica crucible (of about 65-85 ml capacity).

The skidmore crucible is provided with a lid, having a small tube-type opening for the escape of volatile matter. The combination is then placed in a wrought iron crucible (about 8 cm in diameter and 6 mm high) covered with chimney-shaped iron hood (of about 10 cm diameter). The wrought iron crucible is heated slowly for 10 minutes, till flame appears.

Slow heating is continued for 5 minutes more. Finally, strong heating is done for about

15 minutes, till vapours of all volatile matter are burnt completely. Apparatus is then allowed to cool and weight of residue left is determined. The result is expressed as percentage of the original weight of oil taken.

6.3   Cement: Constituents, Manufacturing, Hardening, Setting and Deterioration of Cement

Cement is a hydraulic binder, i.e., an inorganic, non-metallic, finely ground substance which, after mixing with water, sets and hardens independently as a result of chemical reactions with the mixing of water and, after hardening, it retains its strength and stability even under water.

The most important area of application is therefore the production of mortar and concrete, i.e., the bonding of natural or artificial aggregates to form a strong building material which is durable in the face of normal environmental effects.

Portland Cement

$CaO \rightarrow$ 60 to 66%

$SiO_2 \rightarrow$ 17 to 25%

Remaining have small amount like Fe_2O_3, Al_2O_3, MgO, Na_2O & K_2O and SO_3.

Constituents in Cement

- Lime (CaO): It is a very important constituent of cement. The proportion of lime affects the quality of the cement. Presence of excess lime in cement decreases the strength of cement due to the expansion, while lesser proportion of lime reduces the strength of cement due to quick setting. Its preferred percentage is 60-69%.

- Silica (SiO_2) : It is responsible for the strength of the cement. The required percentage of silica is 17-25%.

- Alumina (Al_2O_3) : It is responsible for quick setting of cement, when present in large proportion, can weaken the cement.

- Iron oxide (Fe_2O_3): It gives gray colour and hardness along with silica.

- Sulphur trioxide (SO_3): It should be present in the right proportion to offer soundness (expansivity of cement in 24 hours when the temperature is 25-100°C). If the expansion of the cement during setting is less, then the cement is said to be sound. A large quantity of SO_3 reduces the soundness (soundness is the ability of the cement paste to retain its volume after setting).

- Alkalis: These are responsible for the efflorescent (it is the loss of water of crystallization from a hydrated salt when exposed to air) nature of cement.

- Gypsum: It is used to retard the initial setting of cement.

Manufacturing Process

Portland cement can be made by following two different processes - a dry one and a wet one. The basic ingredients of both the dry and wet processes are the same. By mass, lime and silica make up approximately 85% of Portland cement. The materials that are commonly used are limestone, shells, chalk, shale, clay, slate, silica sand and iron ore.

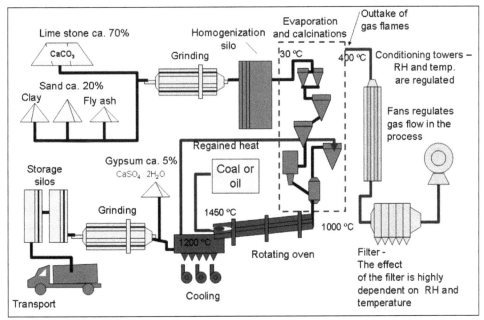

Manufacturing of portland cement.

- Limestone, shale, silica and iron oxides are quarried from the ground. (Some limestones already contain enough silica).

- Rock materials are run through a crusher that turns rock into smaller pieces.

- Crushed limestone + silica + shale + iron oxides are mixed together and run through a rotary kiln.

- Rotary kiln continuously mixes ingredients and "calcines" limestone so that O_2 is driven off, forming clinker.

- Clinker is ground to fine powder and mixed with gypsum then bagged for sale.

Hardening and Setting

The setting and hardening of cement take place between Portland cement and water, i.e., hydration. In order to understand the properties and behavior of cement chemistry of hydration is necessary.

Hydration reactions of the cement are complex that the hydration of each compound takes place independently of the others.

1. Calcium Silicates (C_3S and C_2S)

Hydration of the two calcium silicates give similar chemical products, differing only in the amount of calcium hydroxide formed, the heat released and reaction rate.

$$2C_3S + 7H \rightarrow C_3S_2H_4 + 3CH$$

$$2C_2S + 5H \rightarrow C_3S_2H_4 + CH$$

The hydration product is $C_3S_2H_4$ (calcium silicate hydrate) or C-S-H (non-stoichiometric). This product is not a well-defined compound. The formula $C_3S_2H_4$ is an approximate description. It has amorphous structure making up of poorly organized layers and is called glue gel binder.

C-S-H is believed to be the material governing concrete strength. Another product is $CH - Ca(OH)_2$, calcium hydroxide. This product is a hexagonal crystal often forming stacks of plates. CH can bring the pH value to over 12 and it is good for corrosion protection of steel.

2. Tricalcium Aluminate (C_3A)

Without gypsum, C_3A reacts very rapidly with water:

$$C_3A + 6H \rightarrow C_3AH_6$$

The reaction is so fast that it results in flash set which is the immediate stiffening after mixing, making proper placing, compacting and finishing impossible.

Gypsum the primary initial reaction of C_3A with water is,

$$C_3A + 3(C S H_2) + 26H \rightarrow C_6S_3H_{32}$$

The 6-calcium aluminate trisulfate-32-hydrate is usually called ettringite. Formation of ettringite slows down the hydration of C_3A by creating a diffusion barrier around C_3A. Flash set is thus avoided.

Even with gypsum, the formation of ettringite occurs faster than the hydration of the calcium silicates. It therefore contributes to the initial stiffening, setting and early strength development. In normal cement mixed with ettringite is not stable and will further react to form monosulphate (C_4ASH_{18}).

The rate of hydration during the first few days is in the order of $C_3A \gg C_3S \gg C_4AF \gg C_2S$. Their reactivates can be observed in the following figure.

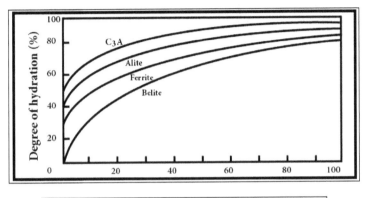

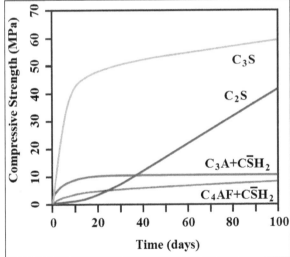

Graph showing the rate of hydration when reactivates.

Deterioration of Cement Concrete

The various reasons for the deterioration of cement concrete are depicted in the figure below,

The various types of deterioration are:

- Deterioration by surface wear.

- Cracking by crystallization of salts in pores in pores.

- Deterioration by Frost Action.

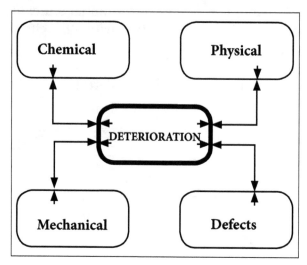

1. Deterioration by surface wear

- Abrasion: Dry attrition (wear on pavements dry attrition and industrial floors by traffic).

- Erosion: Wear produced by abrasive action of fluids containing solid particles in suspension (canal lining, spillways and pipes).

- Cavitation: Loss of mass by the formation of vapor bubbles and their subsequent collapse.

2. Cracking by Crystallization of Salts in Pores in Pores

The crystallization of salts in the pores of concrete can produce stresses that may damage concrete structure.

This can happen when the concentration of the solute (c) exceeds the saturation concentration (C_s). Higher C/C_s ratio (degree of supersaturation supersaturation) produces higher crystallization pressure.

3. Deterioration by the Frost Action

When water freezes, there is an expansion of 9%. However, some of the water may migrate through the boundary, decreasing the hydraulic pressure. Hydraulic pressure depends on:

- Rate at which ice is formed.

- Permeability of the which ice is formed.

- Distance to an "escape boundary".

Aggression by Physical Elements

- Freezing and thawing.

- High temperature.

- Shrinkage and cracking.

Aggression by Mechanical Elements

- Abrasion

- Impact

- Erosion

- Cavitation

6.4   Insulators: Thermal and Electrical Insulators

Insulators

They do not practically allow the heat (or) electricity to flow through them. Example: Most of the organic and Inorganic solids (except graphite).

Electrical Insulator

It must be used in electrical system to prevent unwanted flow of current to the earth from its supporting points. The insulator plays a vital role in the electrical system. It is a very high resistive path through which practically no current can flow.

In the transmission and distribution system, the overhead conductors are generally supported by supporting towers or poles. The towers and poles both are properly grounded. So there must be insulator between tower or pole body and the current carrying conductors to prevent the flow of current from conductor to earth through the grounded supporting towers or poles.

Insulating Material

The main cause of failure of overhead line insulator, is flash over, occurs in between line and earth during the abnormal over voltage in the system. During this flash over, the huge heat produced by arcing, causes puncher in insulator body. Viewing this phenomenon the materials used for electrical insulator has to posses some specific properties.

Properties of Insulating Material

The materials generally used for insulating purpose is called insulating material. For successful utilization, this material should have some specific properties as listed below:

- It must be mechanically strong enough to carry tension and weight of conductors.

- It must have very high dielectric strength to withstand the voltage stresses in High Voltage system.

- It must possess high Insulation Resistance to prevent the leakage current to the earth.

- The insulating material must be free from unwanted impurities.

- It should not be porous.

- There must not be any entrance on the surface of electrical insulator so that the moisture or gases can enter in it.

- There physical as well as electrical properties must be less affected by changing temperature.

Thermal Insulator

Design Criteria and Function

The purpose of thermal insulation in buildings is to maintain a comfortable and hygienic indoor climate at low ambient temperatures. An minimal amount of thermal insulation is required to protect the constructional elements against the thermal impact and moisture related damage.

The main aim of thermal insulation in the winter is energy conservation leading to decrease in heating demand and hence the protection of the environment. This aim has to be considered in new buildings as well as in renovating the building stock. Strategies to reach this aim are the use of building materials with the low thermal conductivity -values and the installation of windows with low U-Values on the on side and avoidance of thermal bridges and uncontrolled infiltration on the other side.

Besides the above mentioned purpose, thermal insulation plays a major role in preventing summertime overheating of buildings through reducing the transmission of the solar radiation, absorbed on the building's exterior surfaces, to the interior. The lowest -values of non-evacuated elements achievable is the one of motionless air.

Hence the basic principle in developing insulation materials is to enclose as much non-moving air into the structure of the material as possible and still satisfy the structural stability required. The lowest achievable -values of insulation elements are realized with evacuated insulation panels.

Material Requirements

Insulation materials have to keep their shape, be durable and resistant against mould and parasites. The moisture balance of construction has to be positive, i.e. the water accumulating inside the construction during thawing period (winter) has to be removed to the exterior during the evaporating period (summer).

The placement of a wind tight layer at outer surface and an airtight on the inner surface prevent the insulation material from being perfused with the humid and warm room air from inside or cold ambient air from outside. Depending on the field of use, insulation materials have to suffice the safety requirements for fire protection.

6.5   Fuel Cells: H_2-O_2 and Methanol Oxygen

Full Cell

A full cell converts the chemical energy of the fuels directly into electricity. The essential process in a fuel cell is,

$$\text{Fuel + Oxygen} \rightarrow \text{Oxidation products + Electricity.}$$

A typical example of pollution free cell is $H_2 - O_2$ fuel cell in which the fuel is hydrogen and the oxidizer is oxygen.

Construction and Working

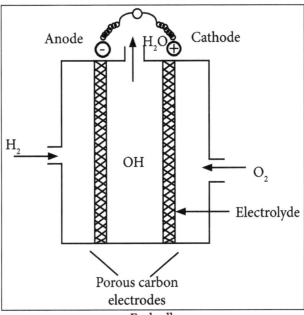

Fuel cell.

The Hydrogen - Oxygen fuel cell consists of two porous electrodes made up of compressed carbon coated with small amount of catalysts (P_t, P_d, A_g) and KOH or NaOH solution as the electrolyte.

During working, hydrogen (the fuel) is bubbled through anode compartment where it is oxidized. The oxygen (oxidizer) is bubbled through the cathode compartment where it is reduced. The following cell reactions occur.

Anodic reaction: $2H_2(g) + O_2(g) \rightarrow 2H_2O(l)^-$

Cathodic reaction: $O_2 + 2H_2O + 4e^- \rightarrow 4OH^-$

Overall cell reaction: $2H_2(g) + O_2(g) \rightarrow 2H_2O(l)^-$

From the above cell reactions, the hydrogen molecules are oxidized to water. When a large number of fuel cells are connected in series it is called as fuel battery.

Advantages of Fuel Cells

- Fuel cells are efficient and take less time for operation.

- No harmful chemicals are produced in the fuel cells.

Uses

- It is used as auxiliary energy source in space vehicles, submarines etc.

- It is used in producing drinking water for astronauts in the space.

H$_2$ -O$_2$ Fuel Cell

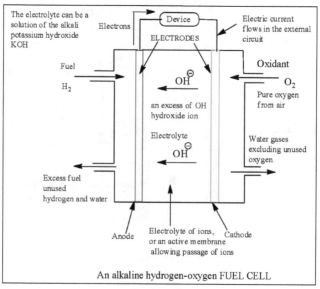

An alkaline hydrogen-oxygen FUEL CELL

H_2-O_2 cell.

A fuel cell is like a battery. It has two electrodes where the reactions take place and an electrolyte which carries the charged particles from one electrode to other. In order for a fuel cell to work, it needs hydrogen (H_2) and oxygen (O_2).

The hydrogen enters the fuel cell at the anode. A chemical reaction strips the hydrogen molecules of their electrons and atoms become ionized to form H^+.

The electrons travel through wires to provide a current to do work. The oxygen enters at the cathode generally from the air. The oxygen picks up the electrons that have completed their circuit. The oxygen then combines with ionized hydrogen atoms (H^+) and water (H_2O) is formed as the waste product which exits the fuel cell. The electrolyte plays an essential role as well.

It only allows the appropriate ions to pass between the anode and the cathode. If other ions were allowed to flow between the anode and cathode, the chemical reactions within the cell would be disrupted.

The reaction in a single fuel cell typically produces only about 0.7 volts that can be used in the cars, generators or other products that require power.

The reactions involved in a fuel cell are as follows:

Anode side (an oxidation reaction): $2H_2 \rightarrow 4H^+ + 4e^-$

Cathode side (a reduction reaction): $O_2 + 4H^+ + 4e^- \rightarrow 2H_2O$

Net reaction: $2H_2 + O_2 \rightarrow 2H_2O$

Applications

The use of fuel cells mainly falls into three broad areas:

- Portable power generation.
- Stationary power generation.
- Power for transportation.

$H_2 - O_2$ fuel cell is used in space crafts, submarines to get electricity. In $H_2 - O_2$ fuel cell, the product water is a valuable source of fresh water for astronauts.

Methanol Oxygen Fuel Cells

In a fuel cell electrical energy is obtained from oxygen and a fuel that can be oxidized. The essential process in a fuel cell is,

Fuel + Oxygen $\rightarrow$ Oxidation product + Electricity

Methanol – oxygen fuel cell are powered by pure methanol. In the early 1960s, the mechanism of this reaction was the subject of many studies in different countries. Only in the mid-1990s, after the great successes achieved in the development of hydrogen-oxygen fuel cells with the proton-conducting ion exchange membrane, was a breakthrough reached in the development of fuel cells with direct (without preliminary conversion) oxidation of methanol.

The fuel cells of this type were now called "direct-methanol fuel cells" (DMFC). The design of modern DMFCs is very similar to the design of proton exchange-membrane fuel cells. The membranes (Nafion) and catalysts (Platinum) are used.

Direct Methanol Fuel Cell (DMFC)

It is a fuel cell that runs directly on methanol (or various liquid fuels) without having to first convert those fuels into hydrogen gas. It consists of two electrodes separated by a proton exchange membrane (PEM) and connected via an external circuit that allows the conversion of free energy from chemical reaction of methanol with the air or oxygen to be directly converted into electrical energy.

Direct methanol fuel cell is a proton exchange membrane (PEM) fuel cell that is fed with an aqueous solution of methanol. The two catalytic electrodes where methanol oxidation (anode) and oxygen reduction (cathode) are separated by a membrane which conducts protons from anode to cathode, while other compounds diffusion is blocked.

The combination of the electrodes and membranes is called membrane electrode assembly (MEA). MEA was prepared with Pt/Ru black at anode and Pt black cathode on either side of a Nafion membrane. Each electrode is made of a gas diffusion layer and a catalytic layer.

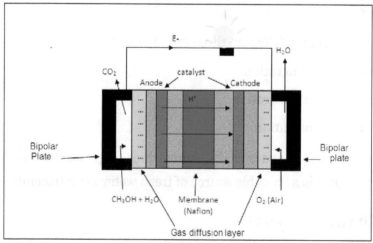

Direct Methanol – Oxygen Fuel Cell (DMFC).

Membrane Nafion was created by addition of sulfonic acid groups into bulk polymer matrix of Teflon. These sites have strong ionic properties and act as proton exchange sites. The aqueous methanol is fed at the anode side.

It diffuses through the diffusion layer to catalytic layer where it is electrochemically oxidized into mainly carbon dioxide, protons and electrons. The protons formed during this reaction diffuse through the Nafion membrane to cathode catalytic layer.

They participate in the oxygen reduction to form water at cathode side. The oxygen may be pure but can also come from air. The electrons are collected by graphite bipolar plates which are the two poles of the cell.

Reactions

Anode reaction: $CH_3OH + 7H_2O \rightarrow CO_2 + 6H_3O^+ + 6e^-$ $E^\circ = 0.02V$

Cathode reaction: $3/2O_2 + 6H_3O^+ + 6e^- \rightarrow 9H_2O$ $E^\circ = 1.23V$

Overall reaction: $CH_3OH + 3/2O_2 \rightarrow CO_2 + 2H_2O$ $E^\circ = 1.21V$

At anode, the methanol is oxidized into carbon dioxide and six protons (as hydronium ions) plus six electrons. The six protons formed react at the cathode with oxygen to form water. The overall reaction looks like a combustion reaction and is thus sometimes referred to as cold combustion.

The cell is a mean to control this reaction and use it to produce current directly. The standard cell voltage for a DMFC at 25°C is 1.21V. However, this potential is never obtained in reality. The open circuit potential is usually about 0.6 to 0.8V.

Issues in DMFC

Slow Electro-oxidation Kinetics

Various surface intermediates are formed during methanol electro-oxidation. The methanol is mainly decomposed to CO which is then further oxidized to CO_2. Other CO like species are also formed: COH ads, HCO ads, HCOO ads Principle by-products are formaldehyde and formic acid.

Some of these intermediates are not readily oxidizable and remain strongly adsorbed to the catalyst surface. Consequently, they prevent fresh methanol molecules from adsorbing and undergoing further reaction.

Thus electro-oxidation of intermediates is the rate limiting step. This poisoning of catalyst surface seriously slows down the oxidation reaction. Besides, a small percentage of intermediates desorbs before being oxidized to CO_2 and hence reduce fuel efficiency but undergoing incomplete oxidation.

Thus, a very important challenge is to develop new electro catalysts that inhibit the poisoning and increase the rate of the reaction. At the same time, they should have a better activity toward carbon dioxide formation.

Methanol Crossover

In DMFC the membrane is partially permeable by methanol dissolved in the aqueous solution. For this reason, some of the methanol penetrates from the anode part of the cell through the membrane into the cathode part. This phenomenon is called as "methanol crossover".

This methanol is directly oxidized on the platinum catalyst by gaseous oxygen without producing useful electrons. This has two consequences:

- A substantial part of methanol is lost for the electrochemical reaction.

- The potential of the oxygen electrode is shifted to lower positive values and therefore the operating voltage of fuel cells is diminished.

In DMFC, the fuel diffuses through Nafion membrane. Methanol that crosses over reacts directly with oxygen at the cathode. Resulting in an internal short circuiting and consequently a loss of current. Besides, the cathode catalyst, which is pure platinum, is fouled by methanol oxidations intermediates similar to anode.

Simple Solutions to Prevent Crossover

Crossover is enhanced by the concentration and pressure gradient between the anode and cathode. It can be easily limited by using a low methanol concentration in the anode feed solution and by increasing the cathode pressure in a certain measure.

A compromise should be found for the concentration. It should be small enough to reduce crossover as much as possible but also supply the anode catalytic layer with enough methanol to produce an acceptable current density.

The effect of a barrier layer was investigated to reduce crossover. Finally, zirconia and silica nano composite membranes were tested instead of Nafion and found to reduce crossover.

Advantages

- The methanol can be refueled quickly and simply.

- The methanol is arguably the most efficient carrier of hydrogen.

- Simplicity of the methanol storage drastically reduces the weight of the system.

- Ideal for low power-density, but high energy-density applications.

Disadvantages

- Fuel can cross electrolyte without reacting, reducing efficiency.

- Low power-density.

- The methanol can corrode various parts of the fuel cell.

- The methanol is toxic and flammable, much like gasoline.

Applications

A potential area of application for DMFCs is low-power (up to 20 W) power sources for electronic equipment, such as notebooks, video cameras, DVD-players, cell phones, medical devices, etc. At this time, the application of DMFCs as power sources for the electric vehicles is very remote.

Substantial progress are also made in the design and prototyping of the DMFC stacks for portable and auxiliary power applications. Three generations of highly efficient and lightweight 11-W stacks were developed for the operation at a design voltage of 0.55/ cell.

Permissions

Index

Printed in the USA
CPSIA information can be obtained
at www.ICGtesting.com
JSHW051352091023
49903JS00006B/123